CRC Series on

COMPUTER-AIDED ENGINEERING

Editor-in-Chief: *Hojjat Adeli*

Neurocomputing for Design Automation
Hojjat Adeli and Hyo Seon Park

High Performance Computing in Structural Engineering
Hojjat Adeli and Roesdiman Soegiarso

Distributed Computer-Aided Engineering for Analysis, Design, and Visualization
Hojjat Adeli and Sanjay Kumar

DISTRIBUTED COMPUTER-AIDED ENGINEERING

for Analysis, Design, and Visualization

Hojjat Adeli
The Ohio State University, U.S.A.

Sanjay Kumar
HyperTree, Inc.

CRC Press

Boca Raton   London   New York   Washington, D.C.

Library of Congress Cataloging-in-Publication Data

Adeli, Hojjat, 1950–
    Distributed computer-aided engineering : for analysis, design, and
visualization / Hojjat Adeli, Sanjay Kumar.
       p. cm.
    Includes bibliographical references and index.
    ISBN 0-8493-2093-3 (alk. paper)
    1. Computer-aided engineering. 2. Electronic data processing-
-Distributed processing. I. Kumar, Sanjay, 1969–   II. Title.
TA345.A328 1998
620′.00285′436—dc21   98-38057
                                                                 CIP

© 1999 by CRC Press LLC

No claim to original U.S. Government works
International Standard Book Number 0-8493-2093-3
Library of Congress Card Number 98-38057
        2 3 4 5 6 7 8 9 0

Preface

Networking of personal computers and workstations is rapidly becoming common place in both academic and industrial environments. A cluster of workstations provides engineers with a familiar and cost-effective environment for high-performance computing. However, workstations often have no dedicated link and communicate slowly on a local area network (LAN) such as ethernet. Thus, to effectively harness the parallel processing or distributed computing capabilities of workstations we need to develop new algorithms with a higher computation-to-communication ratio.

We present a number of distributed algorithms for two fundamental areas of computer-aided engineering: finite element analysis and design optimization. The efficiency and effectiveness of distributed algorithms are demonstrated through examples. The optimization approach used is the biologically inspired genetic algorithm which lends itself to distributed computing effectively.

The distributed algorithms for finite element analysis and design optimization of structures presented in this book provide a new direction in high-performance structural engineering computing. Development of network distributed algorithms can improve the overall productivity of the computer-aided engineering process. They can reduce turnaround time and improve interactivity. Very large models that otherwise cannot fit the local memory of any individual machine can now be analyzed by distribution of data over a group of machines. Further, since workstations are often idle or under-used during nights or weekends, distributed processing can use the idle CPU cycles effectively. As such, we can summarize the motivation for and advantages of distributed computing as

1) Rapid growth in networking of computers,

2) Maximum utilization of available personal computers and workstations, resulting in a substantial reduction of the capital spending requirement for computer hardware, and

3) Provide a relatively inexpensive high-performance computing environment for the solution of large problems that cannot be solved on a single personal computer or workstation within a reasonable time.

Hojjat Adeli and Sanjay Kumar

June 1998

About the authors

Hojjat Adeli received his Ph.D. from Stanford University in 1976. He is currently Professor of Civil and Environmental Engineering and Geodetic Science, Director of the Knowledge Engineering Lab, and a member of the Center for Cognitive Science at The Ohio State University. A contributor to 47 different research journals, he has authored over 300 research and scientific publications in diverse engineering and computer science disciplines. Professor Adeli has authored/co-authored six pioneering books including *Parallel Processing in Structural Engineering*, Elsevier, 1993; *Machine Learning — Neural Networks, Genetic Algorithms, and Fuzzy Systems*, John Wiley, 1995; and *Neurocomputing for Design Automation*, CRC Press, 1998. He has also edited 12 books including *Knowledge Engineering — Volume One Fundamentals*, and *Knowledge Engineering — Volume Two Applications*, McGraw-Hill, 1990; *Parallel Processing in Computational Mechanics*, Marcel Dekker, 1992; *Supercomputing in Engineering Analysis*, Marcel Dekker, 1992; and *Advances in Design Optimization*, Chapman and Hall, 1994. Professor Adeli is the Editor-in-Chief of the research journals *Computer-Aided Civil and Infrastructure Engineering* which he founded in 1986 and *Integrated Computer-Aided Engineering* which he founded in 1993. He is the recipient of numerous academic, research, and leadership awards, honors, and recognition. Most recently, he received The Ohio State University Distinguished Scholar Award in 1998. He is listed in 25 Who's Who and archival biographical listings such as *Two Thousand Notable Americans*, *Five Hundred Leaders of Influence*, and *Two Thousand Outstanding People of the Twentieth Century*. He has been an organizer or a member of the advisory boards of over 130 national and international conferences and a contributor to 103 conferences held in 35 different countries. He was a Keynote/Plenary Lecturer at 24 computing conferences held in 20 different countries. Professor Adeli's research has been recognized and sponsored by government funding agencies such as the National Science Foundation, Federal Highway Administration, and U.S. Air Force Flight Dynamics Lab; professional organizations such as the American Iron and Steel Institute (AISI), the American Institute of Steel Construction (AISC); state agencies such as the Ohio Department of Transportation and the Ohio Department of Development; and private industry such as Cray Research Inc. and Bethlehem Steel Corporation. He is a Fellow of the World Literary Academy and American Society of Civil Engineers.

Sanjay Kumar received his B.E. in 1992 from the University of Roorkee (India) and M.S. in 1994 from The Ohio State University, Columbus, OH where he was a University Fellow and Research Associate under the guidance of Prof. Adeli. His research in the areas of network-distributed algorithms for finite element analysis and genetic algorithm-based structural optimization led to six publications in prestigious international journals. After completing his M.S., Mr. Kumar was employed at Jacobus Technology, Inc. for three years where he was involved in development of distributed object-oriented solutions for engineering automation. He is currently the President of HyperTree, Inc., a consulting firm based in Washington, D.C. area providing distributed object-oriented solutions to Fortune 500 clients such as National Association of Securities Dealers (NASD) and MCI Communications. Besides high-performance computing and distributed-object technology for the internet, his current areas of interest include CORBA, Java, and VRML. He can be reached at sanjay@computer.org.

Acknowledgment

The work presented in this book was partially sponsored by the National Science Foundation, American Iron and Steel Institute, and American Institute of Steel Construction under grants to the senior author. Supercomputing time on the Cray YMP machine was provided by the Ohio Supercomputer Center and use of the CM-5 was provided by the National Center for Supercomputing Applications at the University of Illinois at Urbana-Champaign. Parts of the research presented in this book were published in several journal articles in *Microcomputers in Civil Engineering* (published by Blackwell Publishers), the *Journal of Aerospace Engineering* and the *Journal of Structural Engineering* (published by American Society of Civil Engineers), as noted in the list of references.

Dedicated To

Nahid, Anahita, Amir Kevin, Mona, and Cyrus Dean

Adeli

and

Sitaram Sigh and Uma Devi

Kumar

Table of Contents

1 Introduction

A new trend in high-performance computing in recent years has been development of distributed algorithms on a cluster of workstations using portable message passing libraries such as the Parallel Virtual Machine (PVM). A cluster of workstations provides engineers with a familiar and cost-effective environment for high-performance computing. However, workstations often have no dedicated link and communicate slowly on a local area network (LAN) such as ethernet. Thus, to effectively harness the parallel processing capabilities of workstations we need to develop new algorithms with a higher computation-to-communication ratio.

In Chapter 2 we present distributed algorithms for finite element analysis of large structures on a loosely coupled multicomputer such as a cluster of workstations. A distributed coarse-grained preconditioned conjugate gradient (PCG) solver based on the element-by-element approach is presented to solve the resulting system of linear equations. To account for the slow communication speed of the ethernet network connecting workstations, various techniques are used to coarsen task granularity. A data distribution and data movement strategy is presented based on the set theory. Due to the general nature of the data distribution scheme, the algorithms are versatile and can be applied to analysis of unstructured finite element domains consisting of a combination of various types of elements.

In Chapter 3 implementation of distributed algorithms for finite element analysis of structures on a network of workstations is presented. A common class of object-oriented data structures is created for generation and display of data distribution on the workstations. The performance of distributed algorithms is evaluated using several different types of examples.

In Chapters 4 and 5 we present the computational model for distributed genetic algorithms for optimization of large structures on a cluster of workstations connected via a local area network. The selection of the genetic algorithm is based on its adaptability to a high degree of parallelism. Two different approaches are used to transform the constrained structural optimization problem to an unconstrained optimization problem: penalty function method and augmented Lagrangian approach. For the solution of the resulting simultaneous linear equations the iterative PCG method is used because of its low memory requirement. A dynamic load balancing mechanism is described to account for the unpredictable multi-user, multitasking environment of a networked cluster of workstations, heterogeneity of machines, and indeterminate nature of the iterative PCG equation solver. The implementation of the algorithms using PVM and their application to optimization of large steel structures with hundreds of members are presented in Chapter 5.

In Chapter 6 a mixed computational model is presented for GA-based structural optimization of large structures on massively parallel distributed machines. Parallelism is exploited at both the coarse-grained design optimization level in genetic search using the Multiple Instruction Multiple Data (MIMD) model of computing and the fine-grained fitness function evaluation level using the Single Instruction Multiple Data (SIMD) model of computing. The latter model involves the development of a data parallel iterative PCG algorithm for the solution of the resulting system of linear equations. The model is implemented on the Connection Machine CM-5 and applied to optimization of large space steel structures.

In Chapter 7 we present a computational model for graphic animation of earthquake dynamic response of large structures that can be incorporated into a general-purpose finite element dynamic analysis system in a distributed computing environment. While the CPU-intensive seismic response computations are performed on a supercomputer or a mainframe computer, the animation of a deflected profile of the structure proceeds simultaneously on a workstation. Communication and synchronization constructs from the software library PVM are used for coordination between the heterogeneous applications running on heterogeneous machines. An effective communication strategy for overlapping communications with computations is presented to minimize the effect of network delay. Developed in X Windows, the graphic animation system is based on a hidden line removal scheme combined with an object-oriented knowledge-based system for graphical modeling of the structure.

In the final Chapter 8 we discuss some of the current issues that face computer modelers today in design and development of distributed algorithms for solution of analysis, design, and visualization problems. They include the selection of an appropriate equation solver (direct versus indirect methods), choice of the preconditioner for PCG methods, operating system (Unix versus Windows NT), three-dimensional visualization on the internet, and network infrastructure and Applications Programming Interface (API) for distributed computing.

2 Distributed Finite Element Analysis on a Network of Workstations

2.1 Introduction

In recent years, two developments have had significant impact on finite element (FE) computing. First, the development of adaptive procedures (Rank and Babuska, 1987) driven largely by the growing need for more accuracy in the engineering analysis. Starting from a crude initial mesh just sufficient to model the basic geometry of the domain along with an upper limit of tolerance in accuracy, the adaptive system iterates in a cycle of analysis, error estimation and mesh regeneration until the solution with required accuracy is obtained. Adaptivity introduces a kind of artificial intelligence in the FE analysis process (Adeli, 1988). Users are relieved from the drudgery of complicated mesh generation and data input, thus reducing the costly human involvement in an otherwise computer-automated process. An adaptively refined FE mesh is considerably more detailed than the initial crude mesh and hence requires a greater effort for solution of the resulting system of linear equations. This has led to re-thinking of solution strategies and emergence of iterative solvers such as preconditioned conjugate gradient (PCG) and multigrid methods (Barrett et al., 1993). In these methods, the solution from the previous FE mesh serves as an excellent starting point for the solution of the system of linear equations for the revised mesh, thus leading to a fast convergence to the final solution.

An equally momentous development has been the advent of parallel processing. Faced with the prospect that single-processor performance of a conventional supercomputer has reached its asymptotic limit, engineers have started pursuing the issue of parallelism more aggressively. Examples and surveys of recent work can be found in Adeli et al. (1993), Adeli and Kamal (1993), Adeli (1992a&b), White and Abel (1988), and Noor (1987). For parallel processing, iterative solvers have emerged with significant advantages over conventional direct solvers. Due to a higher degree of concurrency that can be exploited at the element and degree-of-freedom level, iterative solvers are ideally suited for parallelization on parallel supercomputers such as Connection Machine (Johnsson and Mathur, 1989), Hypercube (Malone, 1988), and loosely coupled multicomputers (King and Sonnad, 1987).

Thus, adaptive parallel finite element algorithms based on robust iterative solution methods are likely to become more popular and may ultimately replace the current generation of sequential applications based on direct solvers. However, there is a need for more research in this direction. Bottlenecks exist with regard to both software and hardware. While there have been rapid advances in the development of parallel architectures, most parallel machines are still one-of-a-kind and beyond the reach of small organizations. Also, due to the high cost of application development (contributed to by factors such as non-portability of code and lack of an interactive program development environment), there are few parallel applications in wide commercial use.

A welcome trend in recent years has been toward software environments for development of concurrent algorithms on a cluster of workstations using portable message-passing libraries such as Parallel Virtual Machine (Geist et al., 1993). A cluster of workstations provides engineers with a familiar and cost-effective environment for high-performance computing. However, workstations often have no dedicated link and communicate slowly on a local area network (LAN) such as ethernet. Thus, to effectively harness the parallel processing capabilities of workstations we need to develop new algorithms with a higher computation-to-communication ratio.

We present effective strategies for parallelization of finite element analysis of large structures based on the iterative PCG method for solution of the resulting linear system of equations. Keeping in mind the high cost of communication, the PCG method has been restructured in order to reduce its global communication frequency requirement. Efficient domain decomposition and elimination of the communication requirement through redundant computations are among the other techniques employed. There is no need to assemble the global stiffness matrix, and very large structural models can be analyzed through distribution of data over a network of workstations.

We present a data distribution and data movement strategy based on the set theory. Due to the general nature of the data distribution scheme, the algorithms are versatile and can be applied to the analysis of unstructured FE domains consisting of a mix of various types of elements (such as linear, planar, and solid elements) distributed in any arbitrary topology. This contrasts with previously reported attempts towards distributed FE analysis (Nour-Omid and Park, 1987) where mostly pathological cases of well-structured meshes with simple communication and data-distribution patterns have been studied. In those algorithms little consideration has been given to the management of the communication requirement which rises sharply with the irregularity of the mesh topology. Due to the prohibitive cost of communication on workstation clusters, a reduction in the communication requirement is the single most important performance enhancer for distributed FE analysis.

2.2 FE Analysis and PCG Solution Method

For the linear elastic structural analysis problem, FE modeling requires solution of the system of linear equations:

$$Ku - f = 0 \qquad (2\text{-}1)$$

where f is the external nodal load vector, u is the unknown nodal displacement vector, and K is the global stiffness matrix of the structure given by

$$K = \int_V B^T D B dV, \qquad (2\text{-}2)$$

where D is the stress-strain matrix and B is the linear strain-displacement matrix defined by

$$\sigma = D\varepsilon \qquad (2\text{-}3)$$

$$\varepsilon = Bu \qquad (2\text{-}4)$$

In these equations σ and ε are the vectors of element stresses and strains respectively.

The discretization error $e = \bar{u} - u$ is defined as the difference between the exact and the finite element solution and is expressed in energy norm as (Zienkiewicz and Zhu, 1987)

$$\|e\| = \|\bar{u}\| - \|u\| = \left[\int_V (\bar{\sigma} - \sigma) D^{-1} (\bar{\sigma} - \sigma) dV \right]^{1/2} \quad (2\text{-}5)$$

The energy norm of displacement vector u is defined as

$$\|u\| = (u^T K u)^{1/2} \quad (2\text{-}6)$$

The percentage error in energy norm ς is defined by the expression:

$$\varsigma = \frac{\|e\|}{\|\bar{u}\|} \times 100 \quad (2\text{-}7)$$

Since the exact values of the displacement and stress vectors ($\bar{u}$ and $\bar{\sigma}$) are not known, they are estimated by approximate projection and averaging technique. In this chapter, the least-square stress smoothening technique (Hinton and Campbell, 1974) is used.

2.2.1 Conjugate gradient methods

The major cost of finite element analysis of a large structure is due to the solution of the resulting system of linear equations represented by Eq. (2-1). The discretization level and hence the accuracy of a FE model are often determined by the storage requirement and computational cost of solving the resulting system of linear equations. The most widely used techniques are those based on the Gaussian elimination method, called direct methods. They yield solutions within the bounds of machine accuracy in a certain pre-determined number of operations.

However, direct methods have a serious limitation with regard to memory requirement, since the storage of the assembled global stiffness matrix (K) is required. For large structures this matrix is large and sparse and hence, various sparse matrix storage schemes such as the banded or skyline matrix form are used. But, no matter how efficiently the sparse matrix is stored, its size grows geometrically with the size of the FE model (number of nodes) and may exceed the memory limit of even the largest supercomputers available. Thus, the use of slow out-of-core (secondary) storage becomes unavoidable. In such a situation, iterative solution procedures such as conjugate gradient (CG) methods become an attractive approach because of their low memory requirement. For the iterative solution methods there is no need to assemble the global stuffiness matrix since the computations can be done on an element-by-element basis.

Starting from an initial estimate u^0, the CG method (Jennings and Malik, 1978) produces a sequence of approximate solutions u^k which under the right conditions (symmetric and positive definite K) converges to the desired solution. For any trial solution u^k the associated residual is defined as $r^k = f - Ku^k$ which approaches zero when the solution is reached. To accelerate the convergence, the original system of linear equations is preconditioned by a matrix H and the following preconditioned system of linear equations is solved instead:

$$H^{-1}Ku = H^{-1}f \qquad (2\text{-}8)$$

The modified algorithm is called the preconditioned conjugate gradient (PCG) method. In this chapter, we use $H = W$ where W is a diagonal matrix consisting of the diagonal terms of the K

matrix. This is called the diagonal or Jacobi preconditioner (Golub and Loan, 1989). Table 2-1 presents the details of operations involved in each PCG iteration. The three basic operations in PCG algorithms are matrix-vector product Kp, dot-products, and vector updates (addition or subtraction).

2.2.2 Element-by-element approach

As Table 2-1 shows, the main calculation of a PCG iteration is the matrix-vector product Kp where p is the global direction vector for the search. This step can be performed entirely at the element level without any need to assemble the global stiffness matrix (Carey and Jiang, 1986, Carey et al., 1988, Winget and Hughes, 1985) as

$$Kp = \sum_{e=1}^{N_{EL}} C_e^T k_e C_e p \qquad (2-9)$$

where C_e is an $N_{DE} \times N_D$ Boolean connectivity matrix for element e, N_D is the number of degrees of freedom of the entire structure, N_{DE} is the number of degrees of freedom of an element, k_e is the element stiffness matrix for element e, and N_{EL} is the number of elements. The non-zero terms in C_e correspond to global degrees of freedom that are associated with the local degrees of freedom for a given element. A comparison of memory requirement and computational cost of this approach with the banded storage method is given in Table 2-2.

Table 2-1 Operations in each iteration of the sequential PCG algorithm

Steps	FLOP Count	Type of Operation
Initialize $r^0 = f - Kx^0$	$2N_{EL}N_{DE}^2$ $+ N_D$	Matrix-vector product
$t^0 = H^{-1} \bullet r^0$	N_D	Dot product
$p^0 = t^0$		
Iterate for k = 1, 2,... $h^k = Kp^k$	$2N_{EL}N_{DE}^2$	Matrix-vector product
$n = r^k \bullet t^k$	N_D	Dot product
$d = p^k \bullet h^k$	N_D	Dot product
$\alpha = \dfrac{\alpha_n}{\alpha_d}$	1	
$x^{k+1} = x^k + \alpha p^k$	N_D	Vector update
$r^{k+1} = r^k - \alpha h^k$	N_D	Vector update
$t^{k+1} = H^{-1} \bullet r^{k+1}$	N_D	Dot product

Table 2-1 continued

$\beta_n = h^k \bullet t^{k+1}$	N_D	Dot product
$\beta = -\alpha\dfrac{\beta_n}{\alpha_n}$	2	
$p^{k+1} = t^{k+1} + \beta p^k$	N_D	Vector update
Until $\dfrac{(r^k \bullet t^k)}{(r^0 \bullet t^0)} < Tolerance)$	1	

Table 2-2 Comparison of element-by-element approach with banded storage method

	EBE approach	Banded system
Memory requirement (FP Words)	$\dfrac{N_{EL}N_{DE}(N_{DE}+1)}{2} + 6N_D$	$(b+6)N_D$
Operation count (FLOPS)	$2N_{EL}N_{DE}^2 + 7N_D$	$(2b+6)N_D$

In this table, b is the average semi-bandwidth of the K matrix. The major advantage of the element-by-element (EBE) approach is its much lower storage requirement compared with the banded storage method. In the EBE approach, both computing and memory requirements grow only linearly with the number of elements (N_{EL}).

2.3 Data Distribution for Distributed FE Analysis

The basic idea behind the distributed FE analysis is to decompose the FE mesh into a group of contiguous nodes and elements called subdomains and map one subdomain to a processor. Parallelization of FE analysis based on this approach is presented in Table 2-3. The most important consideration in parallel finite element analysis of large structures is the parallelization of the algorithm for solution of the linear system of equations. Of the three types of computations involved in distributed PCG solution algorithm, vector operations can be parallelized by mapping M/p components of the vectors to each processor, where M is the length of the vector and p is the number of processors. Each processor performs the dot product and vector update operations only on the part of the vectors mapped to it. The matrix-vector product can be performed independently and concurrently for elements in a subdomain except for the elements on the boundary of the subdomain. For such elements a local inter-process communication is required. Dot products require global communication in an all-to-all exchange process.

Partitioning the FE domain into contiguous regions while maintaining optimum balance in the numbers of nodes and elements and reducing the amount of local and global communications are the key to successful distributed finite element analysis.

Once the data distribution and data movement schemes have been developed resolving issues such as 'whom to send/receive from' and 'what to send/receive,' distributed FE analysis can proceed based on cooperation between processors using message-passing for communication.

We present a data distribution scheme based on the set theory. Discussion is general and applies to any finite element mesh consisting of a combination of various types of elements. To describe the underlying ideas clearly we refer to a simple example of a two-dimensional finite element domain consisting of 121 quadrilateral elements and 144 nodes shown in Figures 2-1 and 2-2. In the following sections, we illustrate the rationale behind such a data distribution (primarily to reduce the amount of local and global communications) and describe the concurrent computations in our distributed FE analysis.

2.3.1 Basic definitions

A finite element graph consists of a set Z of N nodes and a set R of N_{EL} elements. The set of global degrees of freedom is called ψ. If all nodes have the same number of degrees of freedom the size of ψ is $N_D = N N_{DN}$, where N_{DN} is the number of degrees of freedom per node. For example, in Figure 2-1 $N_{DN} = 2$ and hence $N_D = 288$. Nodes of an element e are included in the set V_e. The set of degrees of freedom associated with an element e is denoted by ψ_e.

Table 2-3 Steps in parallel FE analysis

Steps in analysis	Comments about parallelization
Read mesh data from disk. Read loading cases and database of mesh connectivities.	Concurrent if disk space is shared through a Network File Server (NFS).
Element stiffness evaluation $$k_e = \int_{V_e} B^T DB dV_e$$	This step is inherently parallel since element-level calculations can be done independently. There is no need for any process communication.
Preconditioning and boundary conditions $$H_{ii} = K_{ii} + C \text{ and}$$ $$f_i = f_i + Cu_i$$ where i is the DOF corresponding to the boundary node in the restrained direction. C is a user-defined large constant and u_i is the prescribed displacement.	Since all elements around a node are included in the same subdomain, the diagonal entries of the global stiffness matrix can be assembled without any inter-process communication.
Solution of linear equations using PCG solver $$Ku = f$$	Each subdomain is mapped to a processor. There is one local inter-process communication and two global inter-process communications in each iteration of the basic EBE-PCG algorithm.

Table 2-3 continued

Steps in analysis	Comments about parallelization
Stress evaluation and least-square projection $$\sigma_e = A^{-1} \int_{V_e} DBu_e dV_e$$ $$A = \int_{V_e} N^T N dV$$	This step is inherently parallel since element-level calculations can be done independently. There is no need for any inter-process communication.
Stress smoothening $$\hat{\sigma} = \sum_{e=1}^{N_{EL}} C_e^T \sigma_e$$ $$(\sigma_i)_{av} = \frac{(\hat{\sigma}_i)}{|E_i|}$$	Smoothening through averaging requires the sum of element-level contributions from all the neighboring elements of a node.
Element-level error estimation $$|e_i| = \int_{V_e} (\sigma_{av} - \sigma) D^{-1} (\sigma_{av} - \sigma) dV_e$$ Global error estimate $$\|e\|^2 = \sum_{i=1}^{N_{EL}} \|e_i\|^2$$	Global sum is found using an all-to-all exchange communication process.

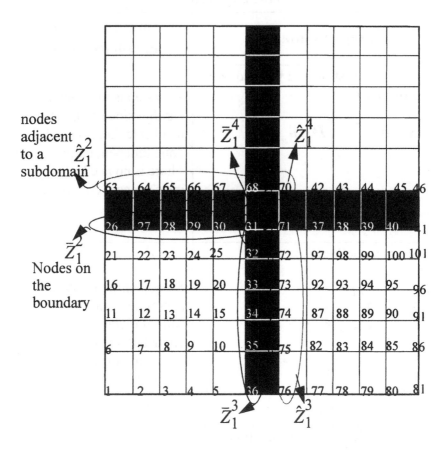

**Figure 2-1. Distribution of nodes among subdomains
(numbers refer to nodes)**

An element is said to be 'adjacent' to a node if the node belongs to its connectivity list. We denote the set of elements adjacent to a node i as E_i. For example, for node 25 in Figure 2-1, the set of adjacent elements is $E_{25} = \{16, 17, 24, 25\}$ (Figure 2-2).

Two nodes are said to be 'adjacent' to each other if they belong to the same element. The set of adjacent nodes of a node i including node i itself, denoted by X_i, is expressed by

$$X_i = \bigcup_{\forall e \in E_i} V_e \qquad (2\text{-}10)$$

For example, for node 31 in Figure 2-1 we can write $X_{31} = \{31, 30, 25, 32, 67, 68, 70, 71, 72\}$. Two elements are adjacent to each other if they share a common node. The set of all elements adjacent to an element e is denoted by L_e. For example, for element 25 in Figure 2-2, $L_{25} = \{16, 17, 24, 29, 30, 31, 32, 33\}$.

2.3.2 Data distribution

A finite element mesh can be divided into N_s subdomains. A subdomain i is characterized by its set of nodes Z_i, their associated degrees of freedom ψ_i, and a set of elements R_i. We impose only one criterion for a valid grouping of elements into a subdomain: a subdomain should include all the elements adjacent to all the nodes in Z_i. Thus,

$$R_i = \bigcup_{\forall j \in Z_i} E_j \qquad (2\text{-}11)$$

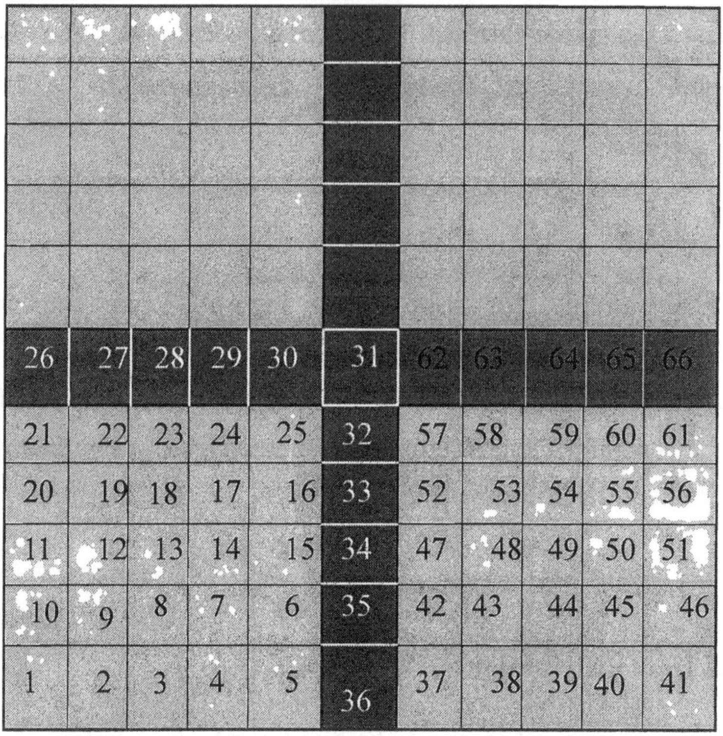

26	27	28	29	30	31	62	63	64	65	66
21	22	23	24	25	32	57	58	59	60	61
20	19	18	17	16	33	52	53	54	55	56
11	12	13	14	15	34	47	48	49	50	51
10	9	8	7	6	35	42	43	44	45	46
1	2	3	4	5	36	37	38	39	40	41

**Figure 2-2. Distribution of elements among subdomains
(numbers refer to elements)**

The elements on the interface of two or more subdomains belong to each one of them, resulting in overlapping of subdomains as far as distribution of elements is concerned (Figure 2-3). The extent of overlap is defined by factor θ:

$$\theta = \frac{\sum_{i=1}^{N_s} |R_i| - N_{EL}}{N_{EL}} \qquad (2\text{-}12)$$

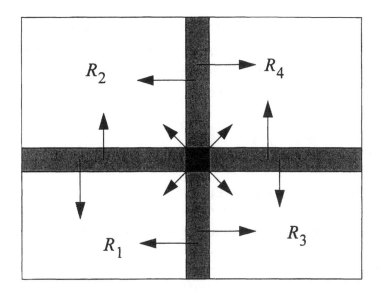

**Figure 2-3. Overlapping nature of subdomains (R_i
refers to set of elements in subdomain i)**

In Eq. (2-12) and subsequent discussions in this chapter the notation $|S|$ refers to the cardinality (number of components) of set S.

The example FE domain shown in Figures 2-1 and 2-2 is divided into 4 subdomains (Figure 2-3). The arrows in Figure 2-3 refer to inclusion of the overlapping interface elements in the subdomains. The sets of nodes in subdomains 1 and 3 are

$Z_1 = \{1, 2, ...36\}$

$Z_3 = \{71, 72, ...76, 77, ..., 101, 37, ..., 41\}$.

The sets of elements in subdomains 1 and 2 are

$R_1 = \{1, 2, 3, 4, \ldots\ldots\ldots 35, 36\}$

$R_3 = \{37, 38, 39, 40, \ldots\ldots, 61, 31, \ldots\ldots, 36, 62\ldots\ldots 66\}$

In this example, the factor θ is computed as 0.193 which means a 19.3% overlap among subdomains. Note that as far as distribution of nodes is concerned subdomains do not overlap and the following relations hold:

$$\bigcup_{i=1}^{N_S} Z_i = Z = \{1, 2, \ldots\ldots\ldots, N\} \qquad (2\text{-}13)$$

$$Z_i \cap Z_j = \varnothing \; i \neq j, \, i, j \in \{1, 2, \ldots\ldots, N_s\} \qquad (2\text{-}14)$$

The corresponding relations for the degrees of freedom are

$$\bigcup_{i=1}^{N_s} \psi_i = \psi = \{1, 2, \ldots\ldots\ldots, N_D\} \qquad (2\text{-}15)$$

$$\psi_i \cap \psi_j = \varnothing \; i \neq j, \, i, j \in \{1, 2, \ldots\ldots, N_s\} \qquad (2\text{-}16)$$

So far, we have not discussed the criteria for grouping of nodes and elements in a subdomain. The simplest approach of course is to include the first $\lceil N/N_s \rceil$ nodes $(1\ldots\lceil N/N_s \rceil)$ in the first subdomain, the next $\lceil N/N_s \rceil$ nodes $(\lceil N/N_s \rceil +1,\ldots.2\lceil N/N_s \rceil)$ in the second subdomain, and so on. (The symbol $\lceil N/N_s \rceil$ refers to the rounded up integer division.) While this simple approach ensures even distribution of nodes, it does not take into consideration the high cost of communication which can result in severe degradation of the performance of the distributed algorithm.

A simple heuristic approach for effective distribution of nodes and reduction of the communication cost (Table 2-4) is summarized as follows:

Step 1: Create a mask vector M containing the number of elements around each node $M(i) = \left|E_i\right|$, i = 1, 2,...., N.

Step 2: Start with a null list of elements and nodes for any subdomain i ($R_i = \varnothing$ and $Z_i = \varnothing$).

Step 3a: Locate a node n with the lowest mask M(n). Add the unassigned elements of E_n, S_1, to subdomain i. Thus, we have $R_i = R_i \cup S_1$.

Step 3b: Add the nodes belonging to the elements in S_1 to subdomain i: $Z_i = Z_i \cup V_j$ $\forall j \in S_1$. Reduce mask M(n) of node n by the number of its adjacent elements assigned to subdomain i, $\left|S_1\right|$.

Step 3c: If current subdomain is full (i.e., $\left|Z_i\right| = \lceil N/N_s \rceil$) $i \leftarrow i + 1$ and start with a new subdomain (i.e., go to step 2). Otherwise, recursively add elements adjacent to elements S_1. Find the list of elements $S_2 = \bigcup_{\forall e \in S_1} L_e$. If $S_2 = \varnothing$ go to step 3a, else $S_1 \leftarrow S_2$ and go to step 3b.

Farhat (1988) presents a domain decomposition and graph coloring method for shared and distributed memory multicomputers which results in subdomains with a balanced number of elements. Adeli and Kamal (1990) present a 3-step substructuring algorithm for frame structures in which both nodes and elements are balanced. Al-Nasra and Nguyen (1992) present certain heuristics which result in closely grouped subdomains eliminating the problem of 'scattering' of subdomains. However, none of these algorithms are directly useful in the context of our distributed FE analysis since they all decompose a FE domain into non-overlapping subdomains.

Due to the overlapping distribution of elements required in the distributed algorithms presented in this chapter, modifications are required in the aforementioned mesh decomposition algorithms. In addition to the three steps previously described in this section, we use an additional step as follows:

Step 4: For all nodes belonging to subdomain i, add their adjacent elements to the list of elements in subdomain i, $R_i = R_i \cup E_j, \quad \forall j \in Z_j$. This step can be added easily to the node-balancing algorithm of Adeli and Kamal (1990).

Note that this will leave the distribution of elements among the processors slightly unbalanced. In our implementation to be presented in the next chapter a class of object-oriented data structures has been developed for partitioning and visualization of the decomposition pattern of FE meshes.

Table 2-4 The domain-decomposition algorithm

Step No.	Steps	Comments
1	$M_i = \lvert E_i \rvert$ $\zeta_e = false$ $\xi_i = false$	Initialize the data structures. The mask of a node M_i is the number of its adjacent elements. ζ_e and ξ_i are boolean toggles to mark node i or element e mapped.
2	$R_i = \varnothing$ $Z_i = \varnothing$ $l_1 = l_2 = l_3 = \varnothing$	Begin a new subdomain. Initialize the lists of elements and nodes to a null set.
3	Find a node n such that $M_n = Min(M_j), j \in \bar{Z}_{i-1}$	Find a node n with the lowest mask on the boundary of the previous subdomain. In the case of the first subdomain, select from the initial list of all nodes (Z).

Table 2-4 continued

Step No.	Steps	Comments

3.1	$\forall e \in E_n$
	
	if (ζ_e = false) then
	
	$\qquad R_i = R_i \cup \{e\}$
	
	$\qquad l_1 = l_1 \cup \{e\}$
	
	$\qquad \forall j \in V_e$
	
	$\qquad\qquad$ if ξ_j = false then
	
	$\qquad\qquad$ Decrement M_j
	
	$\qquad\qquad Z_i = Z_i \cup \{j\}$
	
	$\qquad\qquad$ mark ξ_j = true.
	
	$\qquad\qquad$ Endif
	
	$\qquad$ mark ζ_e = true
	
	Endif
		Add all elements (E_n) around the selected node n and all unmapped nodes belonging to these elements to the current subdomain i and mark the nodes as mapped. Also add the newly added elements to a temporary list l_1.

3.2	if $(|Z_i| >= \lceil N/N_{ss} \rceil)$ then $\qquad i = i + 1$ $\qquad l_2 = l_1$ $\qquad l_3 = \varnothing$	Subdomain i has reached the maximum number of nodes for a subdomain. Go to the next subdomain.

Table 2-4 continued

Step No.	Steps	Comments

4.1	$\forall e \in l_2$
	
	 $\forall k \in L_e$
	
	 if (ζ_k = false) then
	
	 $R_i = R_i \cup \{k\}$
	
	 $l_3 = l_3 \cup \{k\}$
	
	 $\forall m \in V_k$
	
	 if ξ_m = false then
	
	 Decrement M_m
	
	 $Z_i = Z_i \cup \{m\}$
	
	 mark ξ_m = true
	
	 Endif
	 Endif
		For elements in l_2 add recursively all adjacent elements (L_e) and their unmapped nodes (V_k) to the subdomain i and mark the nodes as mapped. Also add the newly added elements to a temporary list l_3.

4.2	if ($|Z_i| \geq \lceil N/N_{ss} \rceil$) then subdomain i is full. $l_2 = \{n \mid n \in l_2 \wedge n \neq e\}$ Goto Step 4.1	Remove the element e from l_2. Restrict the walking-tree-like growth of the subdomain.

Table 2-4 continued

Step No.	Steps	Comments
	$l_2 = l_2 \cup l_3$	Append l_3 to l_2.

4.3	If $l_2 = \varnothing$ then
	
	 Goto Step 3.1
	
	 else
	
	 Goto step 4.1
		Reached a dead end in the heuristic search process. This will result in scattered subdomains. Start with a new node.

4.4	$\bar{Z}_i = \{n \mid n \in Z_i \wedge M_n \neq 0\}$	All nodes with non-zero mask are boundary nodes of current subdomain.

2.3.3 Data movement

For the nodes and elements on the boundary of a subdomain some local communication is required. In this section we identify the data involved in the inter-process communication. Nodes of a subdomain can be classified into two groups: internal and boundary nodes. The set of boundary nodes $\bar{Z}_i$ of a subdomain i is the set of nodes with at least one adjacent node not belonging to the same subdomain i (called non-local nodes). Internal nodes of a subdomain i are denoted by $\tilde{Z}_i$. The set of non-local nodes adjacent to boundary nodes of a subdomain i are called adjacent nodes of subdomain i and are denoted by $\hat{Z}_i$. For example, for subdomain 1 of Figure 2-1, we have

$\bar{Z}_1 = \{26, 27, ...36\}$,

$\tilde{Z}_1 = \{1, 2, ...,25\}$, and

$\hat{Z}_1 = \{63, 64, ...,68,70,.....,76\}$.

Similarly, the components from the set of degrees of freedom ψ are grouped into $\tilde{\psi}_i$, $\bar{\psi}_i$ and $\hat{\psi}_i$ for internal, boundary, and adjacent nodes of subdomain i, respectively. As such, the following relations hold:

$$Z_i = \tilde{Z}_i \cup \bar{Z}_i, \psi_i = \tilde{\psi}_i \cup \bar{\psi}_i \qquad (2\text{-}17)$$

$$Z_i \cap \hat{Z}_i = \emptyset, \psi_i \cap \hat{\psi}_i = \emptyset \qquad (2\text{-}18)$$

The set R_i can be divided into two sub-groups: first is the set of internal elements $\tilde{R}_i$ consisting of elements whose nodes belong to the subdomain i only, and second is the set of boundary elements $\bar{R}_i$ which have at least one non-local node. We have

$$R_i = \tilde{R}_i \cup \bar{R}_i \qquad (2\text{-}19)$$

For example, for subdomains 1 and 3 in Figure 2-2,

$\bar{R}_1 = \{26, 26, 28, ...35, 36\}$,

$\tilde{R}_1 = \{1, 2, 3, 4, 5, ...24, 25\}$,

$$\bar{R}_3 = \{31, ..., 36, 62, 63, ...66\}, \qquad \text{and}$$

$$\tilde{R}_3 = \{37, 38, ...60, 61\}.$$

A subdomain j is 'adjacent' to another subdomain i, if an adjacent node of subdomain i belongs to subdomain j. A_i is the set of all the subdomains adjacent to the subdomain i. The subsets of adjacent nodes of domain i belonging to subdomain j are denoted by $\hat{Z}_i^j$ and the set of boundary nodes of subdomain i adjacent to the subdomain j is denoted by $\bar{Z}_i^j$. For example, for subdomain 1 in Figure 2-1,

$$\hat{Z}_1^2 = \{63, ...,68\},$$

$$\hat{Z}_1^3 = \{71, ...,76\},$$

$$\hat{Z}_1^4 = \{70\},$$

$$\bar{Z}_1^2 = \{26, ...,31\},$$

$$\bar{Z}_1^3 = \{31, ...,36\} \text{ and}$$

$$\bar{Z}_1^4 = \{31\}.$$

The degrees of freedom associated with $\hat{Z}_i^j$ and $\bar{Z}_i^j$ are denoted by $\hat{\psi}_i^j$ and $\bar{\psi}_i^j$, respectively.

2.4 Distributed PCG Algorithm

The most important consideration in development of concurrent algorithms for a particular architecture is task granularity.

Granularity refers to the computation-to-communication ratio. In finite element analysis, we refer to parallelism at the domain level as coarse-grained parallelism. In this case, one complete analysis is performed without any need for inter-process communication. Parallelization at such a coarse-grained level is effective only for some applications. Structural optimization based on genetic algorithms is such an example (Adeli and Cheng, 1994b). We refer to parallelism at the subdomain level as medium-grained parallelism. The distributed FE analysis presented in this chapter and static condensation based on substructuring (Adeli and Kamal, 1990) belong to this category. If there are at least as many processors as the number of elements, an element can be mapped to one processor. This is called fined-grained parallelism where concurrency exists at the element level.

A large amount of communication and redundant computations may be involved in fine-grained parallelization. Thus, fine-grained parallelism can be exploited effectively only on massively parallel architectures with fast interconnection networks such as nCUBE or CM-5. For large-scale FE analysis on workstation clusters, $N_s \ll N_{EL}$ and communication links are slow. Thus, we shall aim for medium-grained parallelization wherein a group of adjacent elements are mapped to one processor. The concurrent element-by-element PCG algorithms are presented in Table 2-5.

Table 2-5 Computations performed by the processor m in the k-th iteration of the parallel element-by-element PCG Algorithms[a]

Basic EBE-PCG	Coarse-grained EBE-PCG
Local communication $$i \in A_m \text{ send}(p_j^k \quad j \in \bar{\psi}_m^i)$$ to processor i and receive(p_j^k $j \in \hat{\psi}_m^i$) from processor i.	Same as basic EBE-PCG algorithm
$$h^k = \sum_{\forall e \in R_m} C_e^T k_e C_e p^k$$	$$^k = \sum_{\forall e \in R_m} C_e^T k_e C_e p^k$$ $$g_i^k = (H_{ii}^k)^{-1} h_i^k$$ $$i \in \psi_m$$
$$\alpha_m^n = \sum_{i \in \psi_m} r_i^k t_i^k$$	$$\alpha_m^n = \sum_{i \in \psi_m} r_i^k t_i^k$$
$$\alpha_m^d = \sum_{i \in \psi_m} p_i^k h_i^k$$	$$\alpha_m^d = \sum_{i \in \psi_m} p_i^k h_i^k$$ $$_m = \sum_{\forall i \in \psi_m} g_i^k h_i^k$$

a. Initialization steps have been excluded for brevity.

Table 2-5 continued

Basic EBE-PCG	Coarse-grained EBE-PCG
Global Sum–Global Broadcast $\forall i \in \{1, 2, ..., N_s\}$, $i \neq m$ Send(α_m^n, α_m^d) to processor i and receive (α_i^n, α_i^d) from processor i.	$\forall i \in \{1, 2, ..., N_s\}$, $i \neq m$ Send(α_m^n, α_m^d, λ_m) to processor i and receive (α_i^n, α_i^d, λ_i) from processor i.
$$\alpha^n = \alpha_m^n + \sum_{i=1,\, i \neq m}^{N_s} \alpha_i^n$$ $$\alpha^d = \alpha_m^d + \sum_{i=1,\, i \neq m}^{N_s} \alpha_i^d$$ $$\alpha = \frac{\alpha^n}{\alpha^d}$$	Same as basic EBE-PCG algorithm.

Table 2-5 continued

Basic EBE-PCG	**Coarse-grained EBE-PCG**
$$x_i^{k+1} = x_i^k + \alpha^k p_i^k$$ $$\forall i \in \psi_m$$ $$r_i^{k+1} = r_i^k - \alpha^k h_i^k$$ $$\forall i \in \psi_m$$ $$t_i^k = H_{ii}^{-1} r_i^k \quad \forall i \in \psi_m$$ $$\lambda_m = -\alpha \sum_{\forall i \in \psi_m^s} h_i^k t_i^{k+1}$$	Same as basic EBE-PCG algorithm.
Global Sum-Global Broadcast $\forall i \in \{1, 2, \ldots\ldots, N_s\}$, and $i \neq m$ send (λ_m) to processor i and receive (λ_i) from processor i. $$\lambda = \lambda_m + \sum_{i=1, i \neq m}^{N_s} \lambda_i$$ $$\beta = \frac{\lambda}{\alpha^n}$$ $$p_i^{k+1} = t_i^{k+1} + \beta p_i^k$$ $$\forall i \in \psi_m$$	$$\lambda = \lambda_m + \sum_{i=1, i \neq m}^{N_s} \lambda_i$$ $$\beta = \frac{\alpha\lambda}{\alpha^d} - 1.0$$ $$p_i^{k+1} = t_i^{k+1} + \beta p_i^k$$ $$\forall i \in \psi_m$$

2.4.1 Concurrent vector operations

Equations (2-13) through (2-16) describe the relationships between nodes and their associated degrees of freedom mapped to different processors. Note that a node belongs to only one sub-domain and is mapped to one processor only. Thus, the vector operations involved in a PCG iteration can be done independently and concurrently on the components mapped to each processor.

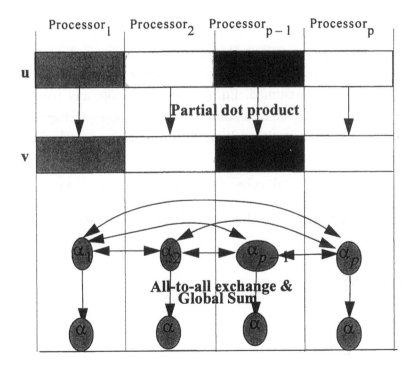

Figure 2-4. Global sum-global broadcast for distributed dot product

For dot products, each processor performs the product operation on the components of the vector mapped to it. Processors broadcast their partial product values to all other processors and receive partial product values from all other processors (Figure 2-4). The sum of the partial product values gives the global value of the dot product. This is a three-step process called global-sum global-broadcast (GS-GB), and is described as follows for two arbitrary vectors u and v:

a) Compute the partial product values:

$$a_i = \sum_{\forall j \in \psi_i} u_j v_j \qquad (2\text{-}20)$$

In actual computations, due to irregular node numbering and unstructured topology of FE meshes, a simple ordering of components cannot be assumed. Eq. (2-20) can handle any irregular mesh automatically since all the scattered components mapped to processor i are correctly multiplied using the proper indexing of the vector.

b) Broadcast a_i to all other processors $\{1,2,....,i\text{-}1, i+1,... N_s)$ and receive the partial values of a_j from processors $j = \{1,2,...$

$N_s \}, j \neq i$

c) Sum the partial values

$$a = a_i + \sum_{j=1}^{N_s} a_j, \, j \neq i \qquad (2\text{-}21)$$

For vector updates, each processor updates the components of the vector mapped to it as follows:

$$u_j = u_j \pm k v_j \quad j \in \psi_i \tag{2-22}$$

Since the results of the vector update operation remain internal to the processor, there is no need for communication.

2.4.2 Concurrent matrix-vector product

For an element-level matrix-vector product (Figure 2-5), first the components of p corresponding to the nodes in the connectivity of the element are gathered from the global array $(p_e = C_e p)$. A local dense matrix-vector product is performed between the element stiffness matrix (k_e) and the direction vector (p_e). Finally, the result is scattered back to the assembled global vector (h).

A degree of freedom (DOF) of the FE model is associated with one row in the system of linear equations. To obtain the result of the matrix-vector product for a row j (corresponding to a degree of freedom j) associated with any node i, contributions of element-level matrix-vector products from all elements adjacent to node i (E_i) are required:

$$v_j = \sum k_e^j C_e^j p, \quad e \in E_i \tag{2-23}$$

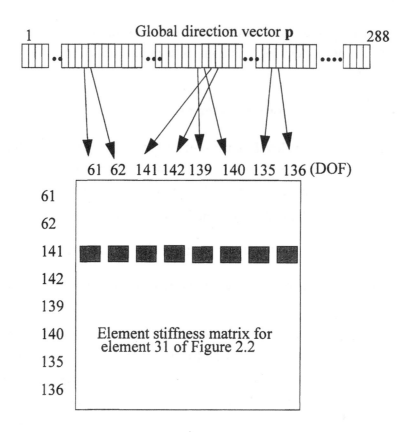

Figure 2-5. Gather-and-scatter operations in element-by-element matrix-vector product

where k_e^j refers to the row of k_e associated with the global degree of freedom j. Figure 2-6 shows rows corresponding to global DOFs 61 and 62 of element stiffness matrices (k_e) for the elements adjacent to node 31 of Figure 2-1. To obtain the matrix-vector product corresponding to DOFs 61 and 62 (corresponding to node 31), the four element-level matrix-vector products should be combined.

For the internal elements $e \in \tilde{R}_i$ of a subdomain i, the matrix-vector product requires no communication since element-level vector p_e can be gathered entirely from the components mapped to the process i (all the nodes V_e are mapped to the processor i). However, for the elements on the boundary of a subdomain $\bar{R}_i$, there is a set of non-local components associated with the non-local nodes $\hat{Z}_i$. Thus, the matrix-vector product for boundary elements requires a local communication for the non-local components ($\hat{\psi}_i$) of the vector p.

A typical local communication involves transfer of components $\hat{\psi}_i^j$ ($j \in A_i$) from the processor j to processor i and a reciprocal transfer of components $\bar{\psi}_i^j$ ($j \in A_i$) from processor i to processor j. Formally, for a processor i, the local interprocess communication can be summarized as

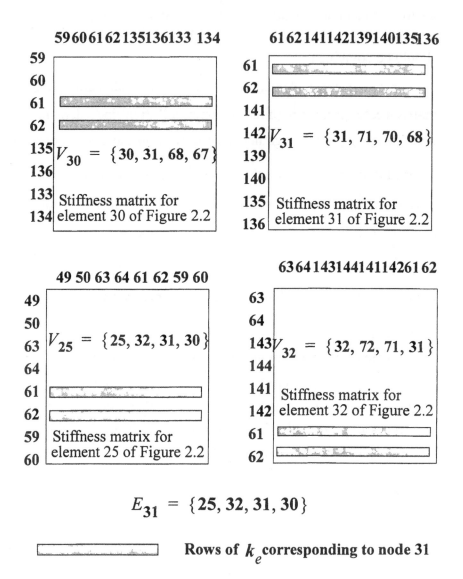

$$E_{31} = \{25, 32, 31, 30\}$$

Rows of k_e corresponding to node 31

Figure 2-6. Element-level rows required for matrix-vector product

$\forall j \in A_i \; Send \, (p_n \; \forall n \in \overline{\psi}_i^j)$ to the processor j and

$\forall j \in A_i \; Receive \, (p_n \; \forall n \in \hat{\psi}_i^j)$ from the processor j.

The total number of local communications is

$$C_n = \sum_{i=1}^{N_s} |A_i| \qquad (2\text{-}24)$$

and the total number of components (FP words) communicated in each PCG iteration is

$$C_{comp} = \sum_{i=1}^{N_s} \sum_{j=1}^{|A_i|} |\hat{Z}_i^j| \times N_{DN} \qquad (2\text{-}25)$$

2.4.3 Redundant computation to reduce communications

For boundary nodes of a subdomain d ($\overline{Z}_d$), the matrix-vector product involves the boundary elements of the subdomain. If an element e in the set E_i ($i \in \overline{Z}_d$) is not included in the subdomain d, there would be a need to communicate the rows of k_e^i from the processor to which e has been mapped. This communication can be avoided by including all the elements around a boundary node in subdomain d. That is, an element is included in all the subdomains to which any of its nodes belongs. For example, in Figure 2-2 element 31 belongs to all the four subdomains. Element-level computations for overlapping elements (such as computation of element stiffness matrices and matrix-vector products) are replicated in all the subdomains. However, the amount of redundant computation due to the overlapping nature of the subdomains is small and is more than compensated for by the resulting saving in communication cost.

2.4.4 Coarsening task granularity

Task granularity can be improved if either the size or number of communications between processors is reduced. We already described in detail the techniques employed to reduce the amount of local communications in the distributed PCG algorithm. In this section we present a modified element-by-element PCG algorithm in which the number of global communications per iteration is reduced by half. This is based on the observation that parameter β (defined in Table 2-1) can be computed in a way that eliminates the need for the second GS-GB step (Table 2-5). We use an alternative expression to compute β, based on the fact that two successive residual vectors r^k and r^{k+1} are orthogonal (Saad, 1985 and Ayakanat et al., 1988):

$$\beta = \alpha\left(\frac{H^{-1}Kp \cdot h}{p \cdot Kp}\right) - 1.0 \qquad (2\text{-}26)$$

We call the modified algorithm coarse-grained EBE-PCG (CG-EBE-PCG) and the modified steps are presented in the right column of Table 2-5. Note that the number of PCG iterations required in both algorithms remains the same.

In the above expression for β, there is an additional dot product involving the diagonal preconditioning matrix H^{-1} and the vector h. Also, at each GS-GB step, three FP words are now communicated instead of the two required in the basic EBE-PCG algorithm. However, the additional time required for transferring one more word and a dot product are negligible compared to the saving in setup time resulting from the reduction in the number of communications.

2.4.5 Memory requirement

The memory requirement of the basic PCG algorithm consists of $6N_D$ words for the storage of five vectors r, x, p, h, t (Table 2-5) and the diagonal terms of the preconditioning matrix H^{-1}, and $N_{EL}N_{DE}(N_{DE}+1)/2$ words for the storage of the dense symmetric element stiffness matrices (k_e). In the distributed PCG algorithms, each processor needs to store only the mapped components of the vectors and the element stiffness matrices of the elements mapped to it. Thus, the memory requirement (in FP words) after distribution of data among the processors is reduced to

$$M_{dist} = \left\lceil \frac{6N_D}{N_s} \right\rceil + \left\lceil \frac{N_{EL}(1+\theta)}{N_s} \right\rceil \times \frac{N_{DE}(N_{DE}+1)}{2} \qquad (2\text{-}27)$$

This substantial reduction in memory requirement facilitates processing of very large FE models on workstations with limited amounts of memory. In addition, a reduction in memory requirement saves time for page swapping to secondary storage of workstations (called virtual memory).

2.5 Final Comments

In this chapter we presented distributed algorithms for finite element analysis on a cluster of loosely coupled computers. The

algorithms are based on and can be implemented on any distributed memory architecture. The major concern for such an architecture is the cost of communication. Effective strategies for
reducing communication cost for an arbitrary FE domain were
presented. The implementation and application of the algorithms
are described in the next chapter.

3 Implementation of Distributed Algorithms for Finite Element Analysis on a Network of Workstations

3.1 Introduction

In Chapter 2 we presented distributed algorithms for finite element analysis of large structures. In this chapter we present the implementation and performance evaluation of the algorithms on a network of workstations. Key to the success of the next generation of FE analysis software with advanced capabilities such as parallel processing is the development of interactive user interfaces, graphical pre- and post-processing and performance monitoring tools. In this book, a common class of object-oriented data structures has been developed for generation and display of data distribution among the workstations and inter-process communication scheme. The resulting algorithms are robust and versatile and can process FE models consisting of a mix of various types of elements distributed in any arbitrary topology.

The main steps in distributed FE analysis are concurrent I/O, concurrent generation of data distribution and data movement scheme, concurrent element stiffness matrix evaluation, and distributed preconditioned conjugate gradient (PCG) iterations followed by post-processing and error estimation. First, the basic model data are read from the hard disk. Since disk space of workstations are shared through a network file server, reading of data can proceed concurrently. Subsequently, all workstations perform domain decomposition independently and obtain the data distribution and data movement scheme for the subdomain mapped to them. This small step is duplicated in all the workstations. Next, each workstation performs an element stiffness matrix evaluation and proceeds to the PCG iterations. Iterations of element-by-element PCG (EBE-PCG) algorithms require local and global communications and synchronization between the workstations. Message-passing constructs from the software library PVM (Geist et al., 1993) are used for communication between workstations. After each iteration, a test of convergence is performed simultaneously by each workstation.

The programming model used in this chapter is called Single Program Multiple Data model (SPMD). In this model all processes are identical during most of the computations and work on different sets of data concurrently. This is a preferable style of programming for distributed systems since the user has to maintain only one source code. The SPMD model is suitable when a large amount of data needs to be processed concurrently. For such applications, this is a more scalable approach than the client-server or master-slave model where one master workstation controls a number of slave workstations.

3.2 Object-Oriented Data Structures

It is interesting to observe that the data structure required for the domain decomposition algorithm presented in Chapter 2 is the same as that required in graphical pre- and post-processing of finite element models. Thus, a common class of data structures is developed to accomplish both tasks of data distribution and graphical display of distribution pattern. The various lists of connectivities existing in the finite element model should be easily and quickly retrievable from memory. In our implementation, the object-oriented programming paradigm has been used for this purpose (Yu and Adeli, 1993; Adeli and Yu, 1995; Adeli and Kao, 1996; Kao and Adeli, 1997).

Object-oriented (OO) programming is a programming approach based on the concepts of data abstraction, inheritance and polymorphism. Based on these concepts, OO programming allows more efficient reuse, maintenance and modification of existing code. The basic building block of object-oriented programming (OOP), object, is an integrated unit of data and methods (also called procedures or functions) acting on these data. Objects are abstracted as classes. A class is a description of behavior of a group of objects (the instances of the class). It describes the nature of their internal data and the methods that are executed in response to messages. Thus, class is a blue print or template for the creation and destruction of its instance objects.

Inheritance is the ability of data types to inherit data elements and operations from other data types. This OOP feature is used to express special relationships between classes and to arrange them into hierarchies. The class at the top of the inheritance hierarchy is called *base class* and the new classes that inherit properties from the base class are called *derived classes*. In the following section we explain the function of a sample *Element* class.

3.2.1 Modeling the hierarchy of elements

In finite element analysis, all elements have certain common features: they all have a list of nodes associated with them. Thus, all common features of elements are encapsulated in a superclass or base class called *Element*. Based on their common properties, elements can be grouped further into families, introducing a second layer in the hierarchy (derived from the superclass *Element*). For example, the popular iso-parametric quadrilateral elements form a family of elements with similar shape functions and stiffness matrix evaluation procedures. We can then write a general element stiffness matrix evaluation procedure for this family of iso-parametric elements. In the example of Figure 3-1, the first layer of derived classes are *Wedge, Quadrilateral, Triangle, Brick, Tetrahedra, Skeletal.* Different element types have many features which are specific to them. A second layer of derived classes (third in the hierarchy) takes care of individual characteristics of different element types. Various types of element supported by our distributed finite element analysis system are presented in Figure 3-2.

Extensibility of this model can be observed by one simple example: addition of a new element type. It only requires the addition of a derived class to the superclass (abstracting a more general family of elements). For example, to add a cubic brick element type, we need to derive the class *CubicBrick* from the class *Brick*. The new class *CubicBrick* inherits all the methods of its parent class. This means we need to define only the specific aspects of the new element type (such as its shape function).

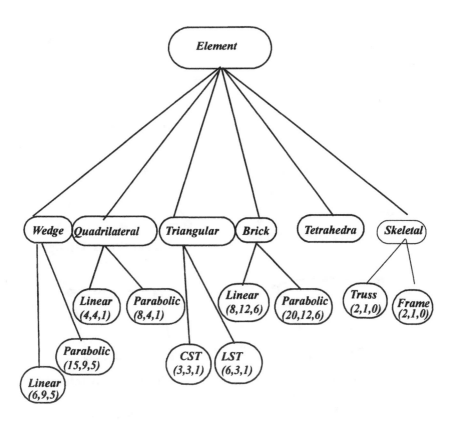

CST - *Constant strain triangle*
LST - *Linear strain triangle*

Numbers in parentheses refer to the number of nodes, edges, and surfaces in each element. Parabolic elements have a mid-side node on the edges.

Figure 3-1. The hierarchy of elements

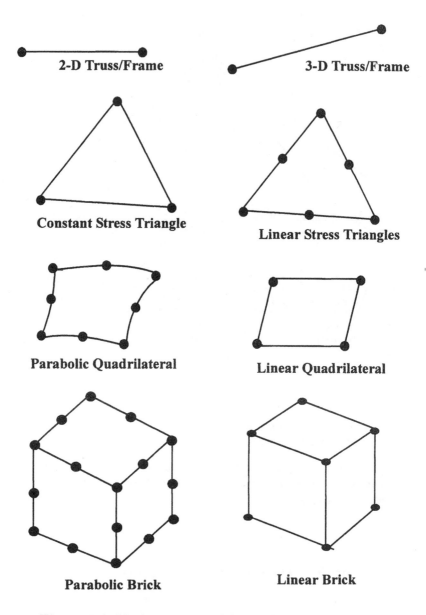

2-D Truss/Frame

3-D Truss/Frame

Constant Stress Triangle

Linear Stress Triangles

Parabolic Quadrilateral

Linear Quadrilateral

Parabolic Brick

Linear Brick

Figure 3-2. Various types of finite elements supported by the distributed finite element analysis system

3.2.2 Knowledge base of connectivities

Other components of finite element mesh such as nodes, edges, and surfaces are each represented through instances (objects) of classes *Node*, *Edge*, and *Surface*. For each of these classes we define a corresponding *container* class. For example, *ElementContainer* contains information like the number of elements, their common salient features such as dimension (2-D or 3-D), the degree of freedom per node, and the maximum number of nodes per element. They also hold important common properties of objects and provide methods for handling disk I/O (related to each object). This two-level organization is useful in maintaining the dynamic nature of the knowledge base.

Once the basic components of knowledge base (KB) such as *Element* and *Node* have been read from the disk, the list of connectivities is generated through simple message-passing between the objects. This knowledge base is created from input of element connectivities with a small processing requirement. Further, the KB needs to be generated only once and reused repeatedly in many iterations of analysis as long the basic connectivities remain the same. For example, the same KB can be used for distributed FE analysis on a variable number of workstations.

The main list of connectivities in a finite element mesh are:

-nodes of all elements
-elements adjacent to each element
-elements adjacent to each node
-surfaces of each element
-edges of each element
-elements attached to an edge
-elements attached to a surface
-nodes adjacent to a node
-mid-nodes on the edges of elements
-edges of surfaces (for three-dimensional solid elements)

The average and maximum numbers of components in the aforementioned list give a good measure of the topology and density of the mesh. These numbers indicate to some extent the growth in communication requirement as the number of workstations is increased. Thus, this information is useful in domain-decomposition, and front-width and band-width minimization routines. In the subsequent section we outline a procedure to obtain some of the basic data (pertaining to distributed finite element analysis) to create aforementioned lists of connectivity for complex mesh structures.

3.2.3 Creating a list of adjacent nodes of a node

Two nodes are adjacent if they belong to at least one element in common. If Node 1 is adjacent to Node 2, then Node 2 is adjacent to Node 1. Further, a node is adjacent to itself. In Figure 3-3 (consisting of 27 2-D truss/frame elements) the adjacency list of node 8 is {7, 10, 11, 12, 8, 9, 6, 5, 4} of cardinality nine. Similarly, the cardinality of the adjacency list of node 11 is six.

Pseudo-code for creating the list of adjacent nodes of all nodes is given in Table 3-1. In this table a -> denotes a message being sent to the class on the left of the -> sign. At the end of the

Table 3-1 Pseudo-code to create a list of adjacent nodes of all nodes

```
ElementContainer::Create Node AdjacencyList {

Loop [i] over all elements

 Loop [j] over all its nodes

 Loop [k] over all its nodes

   if( j ≠ k)

   NodeContainer ->AddNodeToNode( NodeList[j], NodeList[k]

}

NodeContainer:: AddNodeToNode(node N1, node N2)

With Node N1 do

  if N2 is not already in its adjacency list

    add N2 to the list of adjacent nodes

    increment no of adjacent nodes by 1

}
```

message-passing scheme outlined in Table 3-1 we will be left with an adjacency list of all nodes in the individual instance of the *Node* class. A list of all the nodes in the model is stored by an instance of *NodeContainer* class. The nodal connectivity data is stored as part of individual objects of the *Node* class. The *Node* class has public methods to access the adjacency list.

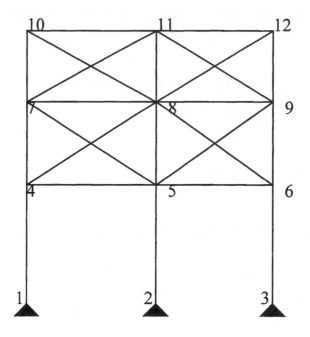

Figure 3-3. Example illustrating adjacent list of nodes

An implementation of the above message-passing scheme is straightforward in an object-oriented language like C++. In general in an OO system, we write several small functions to implement parts of a task in modular fashion, rather than writing one large routine. This ensures better encapsulation and maintains locality of reference. A collection of small, coordinating tasks is as powerful as any large routine and provides more flexibility for further improvement and code optimization.

3.2.4 List of adjacent elements of an element

Two elements are adjacent to each other if they share at least one node. This information is often required in domain decomposition and graphical pre-sorting for hidden line removal. Pseudocode for creating this list is given in Table 3-2.

In Figure 3-4 (consisting of 11 3-D linear brick elements), the adjacency list of element 3 is $\{1, 2, 4, 5, 6, 11\}$. Of these elements, $\{2, 5\}$ are edge neighbors and $\{1, 4, 6\}$ are face neighbors. Also, element $\{11\}$ is only a node neighbor. In general, all surface neighbors are edge neighbors and all edge neighbors are node neighbors.

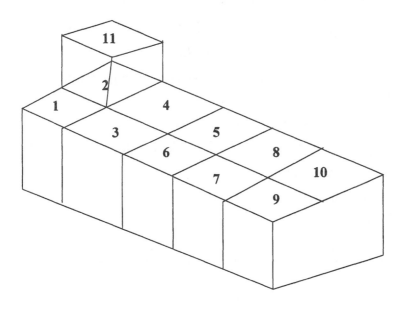

Figure 3-4. Example illustrating adjacent list of elements

Table 3-2 Pseudo-code to form element adjacency list of all elements

Loop [i] over all elements

Loop [j] over all its nodes

  Loop [k] over all elements adjacent to this node

    if element[l] is already in the list of node neighbor

      make it an edge neighbor

    else if element[l] is an edge neighbor do nothing

    else mark this element as node neighbor

  End loop k

End Loop [j]

Loop [j] over all internal edges

  Loop [k] over elements adjacent to this edge

  if element[l] has been marked as a node neighbor then

    make it an edge-neighbor

  else if element[l] has been marked as edge neighbor then

    make it a surface neighbor

  else if element[l] has been marked as surface neighbor

    do nothing

  end loop k

end loop j

End loop i

3.2.5 Domain decomposition

Once the knowledge base of connectivities has been created it can be interfaced to the domain decomposition algorithm

(Section 2.3). The results from the domain decomposition for every subdomain i include:

a) a list of its elements R_i (grouped further into internal elements $\tilde{R}_i$ and boundary elements $\bar{R}_i$);

b) a list of its nodes Z_i (grouped further into internal nodes $\tilde{Z}_i$ and boundary nodes $\bar{Z}_i$);

c) a list of its adjacent subdomains (A_i); and

d) a list of boundary nodes for which the components of global direction vector (p) need to be received $(\hat{Z}_i^j)$ from every adjacent subdomain j, and a list of boundary nodes for which the components of the global direction vector need to be sent $(\bar{Z}_i^j)$ to every adjacent subdomain j.

3.3 Graphic Visualization Using Hidden Line Removal

Based on information stored in the knowledge base of connectivity, the data distribution pattern can be visualized using a hidden line removal scheme. Data distribution can be displayed graphically in an interactive environment. Solid models are passed through a hidden-line filter to minimize image cluttering and improve picture quality. The net result is that all the elements belonging to the same subdomain are shown in the same color. This is helpful in the development of the domain decomposition algorithm itself. If the domain decomposition results in excessive scattering of subdomains, the algorithm needs to be improved. In our implementation, X window system (Chuang and Adeli, 1993), a network-transparent graphic window system for workstations, has been used to display the resulting image of the decomposed FE mesh.

Edges, surfaces, and nodes that do not belong to any of the selected elements or are completely inside the model (and hence are invisible) are filtered out through a simple one-pass search through the knowledge base.

Also, duplicated edges and surfaces (which are completely inside the model) are removed through intelligent pre-sorting. All edges that have more than four elements adjacent to them are removed. Similarly, all surfaces which have more than one element adjacent to them are removed from the list of selected surfaces.

Efficient pre-selection of edges and surfaces reduces the computational cost of hidden line removal substantially since the number of steps in hidden line removal is directly proportional to the number of edges times the number of surfaces. A simple example from the Taj Mahal structure, to be described in Section 3.4.3, illustrates the point (Table 3-3). As the table shows, the actual number of selected surfaces is approximately half of the total number of surfaces. Similarly, approximately one-third of the edges are selected. This means a gain in speed of graphical processing by a factor of six.

After the efficient pre-sorting and selection of edges and surfaces, we eliminate hidden lines and surfaces using a hidden line removal algorithm.

**Table 3-3 Components of knowledge base for the Taj Mahal
structure (Figures 3-14 to 3-16)**

	Starting number	Selected (after pre-sorting)	After removing duplicates	Visible (after hidden line removal)
Elements	831	831	183	126
Nodes	1324	1324	396	177
Edges	9972	3496	812	308
Surfaces	4986	2994	418	130

3.3.1 Hidden line removal

A hidden line removal scheme (Janssen, 1983) filters out all
the invisible edges (which are hidden by any surface in front of it)
in a 3-D model. First, an edge is broken into a number of lines (an
edge can have an arbitrary number of mid-nodes) and a surface is
reduced to a number of facets (3-point planes) by breaking it into
triangles. The 3-D model is now reduced to a collection of lines
and triangles. Next, graphical transformation is applied to the
nodes to get perspective projection of the model on the desired
view-plane. The user has the option to select any angle and distance to view the model. The depth cue (the Z coordinate obtained
after applying the transformation) of the nodes is preserved to
find out the visibility of each line against every plane in the
model. Depending on the depth cue of the two end points of the
line and its intersection with the plane of the triangle, we estimate
if the entire line (or a part of it) is hidden by the triangle. This

gives rise to eleven different possibilities which are presented in Table 3-4.

Output from the hidden line removal algorithm can be searched to find the visibility of surfaces. Visible surfaces of elements are filled with color representing the color of the subdomain to which they belong. Since a visible surface can have only one element associated with it, the subdomain number of the element is used to decide the color of the surface. If an element is on the boundary of a subdomain, it belongs to multiple subdomains. In that case, all the surfaces of such elements are displayed on a common (black) color. To display the result of data distribution and hidden line removal in an interactive environment, X window system, the industry standard for vendor independent graphics on workstations, has been used.

3.4 Examples and Results

We evaluate the performance of the distributed FE algorithms by presenting three examples: a plane stress model, a three-dimensional (3-D) highrise building structure, and a 3-D solid model of a monumental building. The dedicated workstation cluster used consists of six IBM RS/6000 Model 320H machines. The algorithms have been implemented in C++. The tolerance for convergence of the PCG iterations is set at 1.0×10^{-5} for all the examples.

Table 3-4 Hidden line removal algorithm

Case	Figure	Logic used to decide if the line is hidden by a triangle
1 and 2. Both points are outside the triangle		If the line intersects two sides of the triangle, compare the depth cue (z-coordinate) at either point of intersection. If the line is behind the triangle at the point of intersection, then the line has a hidden segment.
		If the line does not intersect the triangle, then the line is not hidden by this triangle.
3. One point is inside and the other point is outside.		Compare the depth cue at the point of intersection (point 3 in this figure). If the line is behind the plane at this point, line segment (1,3) is hidden.
4. Both points are inside.		Compare the depth cue at any of the end-points of the line. If the end point of the line is behind the plane, then the line is completely hidden by this plane. Once a line is marked as entirely invisible, stop processing this line.

Table 3-4 continued

Case	Figure	Logic used to decide if the line is hidden by a triangle
5, 6, and 7. One point is on the edge		If the other point, 2, is inside the plane, then compare the depth cue of point 2 with that of the plane. If the point 2 is behind the plane, the line is marked as completely hidden. Stop processing this line against other surfaces in the model. If the line intersects an edge of the plane (at point 3 in this figure), then compare the depth-cue at the point of intersection. If point 3 is behind the plane, then section (1,3) of the line is hidden. If the other point of the line is outside the plane, then the line is not hidden by this triangle.

Table 3-4 continued

Case	Figure	Logic used to decide if the line is hidden by a triangle
8,9, and 10. One point is on the corner of the plane		If the other point is outside the plane and the line intersects with any one side of the plane (point 2 in this figure), then compare the depth cue at the point of intersection. If point 2 is behind the plane, then section (1,2) of the line is hidden. If the other point is inside the plane, the line is completely hidden. If the other point is outside the plane, then the line is not hidden by this triangle.
11. Numerous other trivial cases.		If both points of the line coincide with one or two different nodes of the plane, then the line is not hidden by this triangle.

3.4.1 Example 1: Plane stress problem

The plane stress L-shaped domain (Figure 3-5) has been discretized using four different types of elements: parabolic quadrilateral (Example 1a), linear quadrilateral elements (Example 1b), parabolic quadrilateral (Example 1c) and linear triangular (Example 1d) elements. The FE domain is modeled as three identical square blocks each discretized identically into 35 subdivisions in both directions. The numbers of nodes and elements generated vary according to the type of the elements used. A summary of discretization data for various types of elements is given in Table 3-5. The finite element mesh using quadrilateral (linear or parabolic) elements decomposed into four subdomains is shown in Figure 3-6. The black regions between the subdomains show the overlapping regions. Loading consists of a distributed load of intensity 1 on the left edge. The edges on the top and right end are restrained by roller supports. The modulus of elasticity is 1.0×10^5 and Poisson's ratio is 0.3. The percentage error in energy norm and number of PCG iterations for the examples is also given in Table 3-5. As expected, parabolic elements are more accurate (as indicated by smaller % errors in the energy norm). However, on parallel machines, the efficiency of parallelization should also be considered in selecting the right type of element.

Variation of elapsed time (also called wall-clock time) per PCG iteration for Examples 1a and 1c (parabolic quadrilateral and triangular elements, respectively) is given in Figure 3-7 and that for Examples 1b and 1d is given in Figure 3-8 for both the coarse-grained (CG) and the basic EBE-PCG algorithms.

Table 3-5 Summary of discretization data for plane stress problem

Example	Type of elements	Number of elements	Number of nodes	% error in energy norm	Number of PCG iterations
1a	Parabolic quadrilateral	3675	11306	3.49	1415
1b	Linear quadrilateral	3675	3816	4.67	492
1c	Parabolic triangle	7350	14981	3.48	1592
1d	Linear triangle	7350	3816	6.50	655

Comparison of coarse-grained PCG and basic PCG algorithms in terms of elapsed time per iteration shows an interesting pattern. For one or two workstations the elapsed time per iteration is higher for the CG-EBE-PCG algorithm and the savings in communication costs do not offset the additional computational cost of one additional dot product. This trend reverses when the number of workstations is increased to 3 or 4. The coarse-grained PCG algorithm emerges as a more efficient algorithm as the number of workstations is increased.

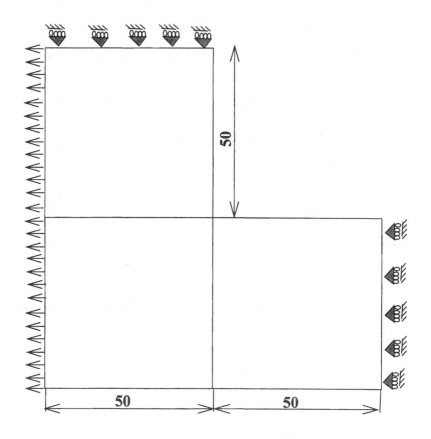

Figure 3-5. Example 1 (Plane stress problem)

The variation of parallel speedup for Examples 1a and 1c is shown in Figure 3-9 and that for Examples 1b and 1d is shown in Figure 3-10. A higher speedup is obtained for the coarse-grained PCG method compared with the basic PCG algorithm for a larger number of machines. As Table 3-6 shows, the amounts of communication and redundant computations (as indicated by the percentage of overlap) involved in Example 1a are less than those for Example 1c. Since floating point operations (FLOPs) per iteration are nearly the same for both examples, the computation-to-communication ratio is higher for Example 1a. Thus, higher speedup is observed for Example 1a compared with Example 1c (Figure 3-9).

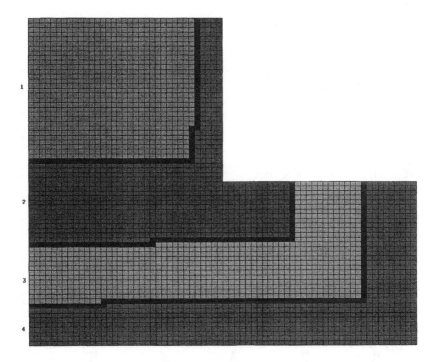

Figure 3-6. A finite element mesh for Example 1 with four subdomains

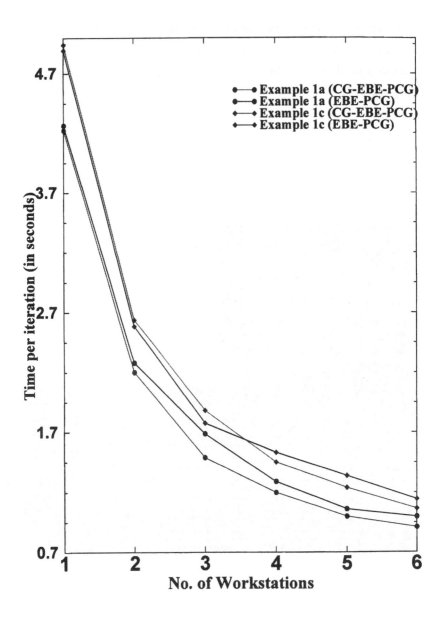

Figure 3-7. Elapsed time per iteration for Examples 1a and 1c

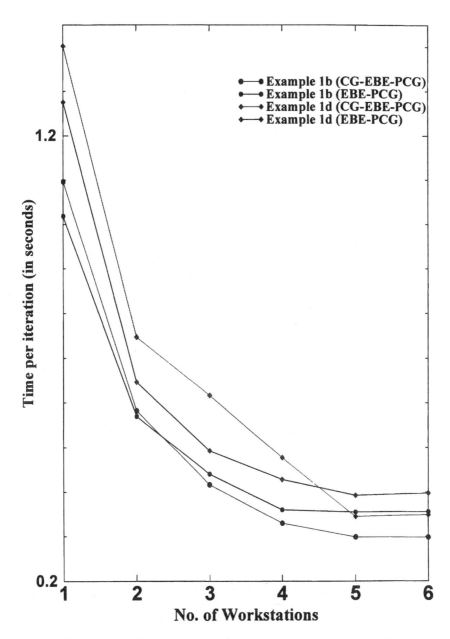

Figure 3-8. Elapsed time per iteration for Examples 1b and 1d

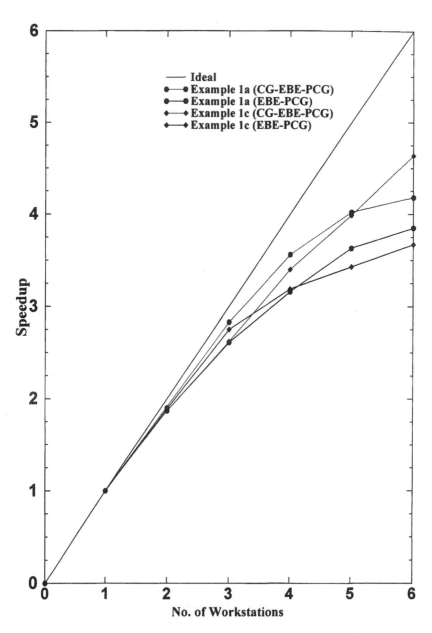

Figure 3-9. Parallel speedup for Examples 1a and 1c

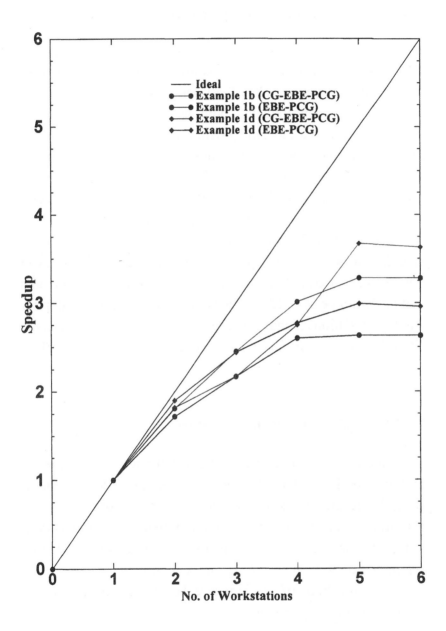

Figure 3-10. Parallel speedup for Examples 1b and 1d

3.4.2 Example 2: 100-story high rise building structure

Example 2 is a 100-story space structure (Figures 3-11 and 3-12). It is intended to model the exterior envelope of a highrise steel building structure. It has 5980 elements and 1240 nodes. The structure has been modeled both as an axial-load structure (Example 2a) and a moment-resisting frame structure (Example 2b). In the former case, the element stiffness matrix has a dimension of 6x6 and there are three degrees of freedom per node resulting in 3720 global degrees of freedom. For the moment-resisting frame structure, the element stiffness matrix has a dimension of 12x12 and there are six degrees of freedom per node resulting in 7440 global degrees of freedom. The number of PCG iterations required in both algorithms for Example 2a is 486 and that for Example 2b is 1241. The variation of parallel speedup for 2 to 6 workstations is shown in Figure 3-13. It is observed that the CG-EBE-PCG algorithm is more efficient than the basic EBE-PCG algorithm.

3.4.3 Example 3: Taj Mahal structure

Example 3 is a solid FE model of a complicated monumental building, the Taj Mahal. This model has been selected to demonstrate the generality and versatility of the domain decomposition algorithm in handling complex three-dimensional meshes consisting of various types of elements. The structure is a doubly symmetric double dome resting on eight columns (Figure 3-14). Only one quadrant of the structure has been modeled taking advantage of the symmetry.

Table 3-6 Summary of domain decomposition for the example structures using four workstations

	Plane stress problem (Example 1a)	Plane stress problem (Example 1c)	100-Story structure (Example 2)	Taj Mahal Structure (Example 3)
% of interface nodes	8.0	7.5	9.2	69.6
Percentage of overlapping elements θ	4.9	5.1	4.5	38.6
Total number of local communications	6	6	6	6
Average number of nodes involved in a local communication	150.6	176.5	17.2	154.5
Number of elements/ Nodes in subdomain 1	958 / 2827	1944 / 3747	1518 / 310	293 / 332
Number of elements/ Node in subdomain 2	1004 / 2826	1980 / 3746	1652 / 309	311 / 330
Number of elements/ Node in subdomain 3	967 / 2826	1949 / 3743	1593 / 309	307 / 330
Number of elements/ Node in subdomain 4	925 / 2827	1849 / 3745	1484 / 312	241 / 332

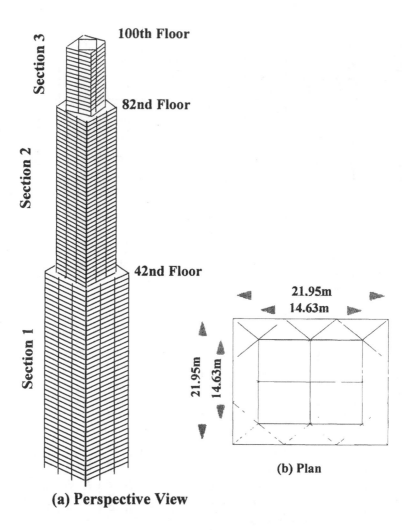

(a) Perspective View

(b) Plan

Figure 3-11. Example 2 (100-story highrise building structure)

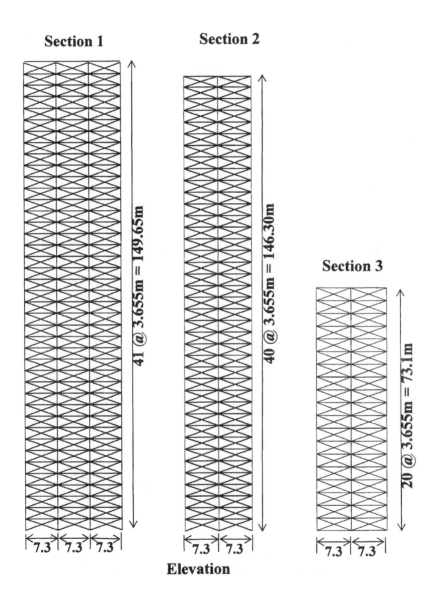

**Figure 3-12. Side view of the 100-story highrise
building structure**

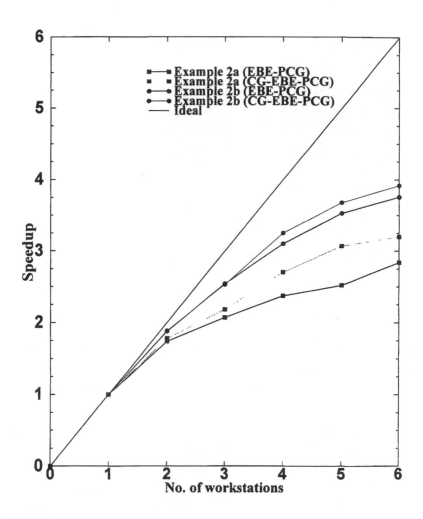

Figure 3-13. Parallel speedup for Example 2

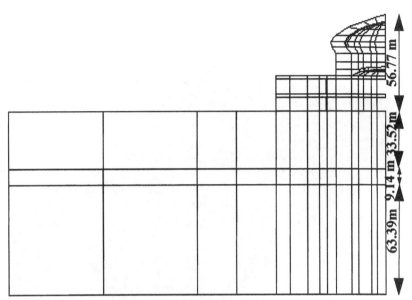

Cross-sectional elevation of a quadrant

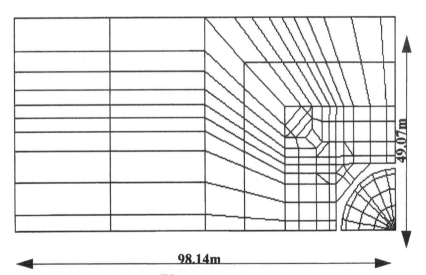

Plan
Figure 3-14. Plan and cross-sectional elevation of a
quadrant of Taj Mahal structure (Example
3)

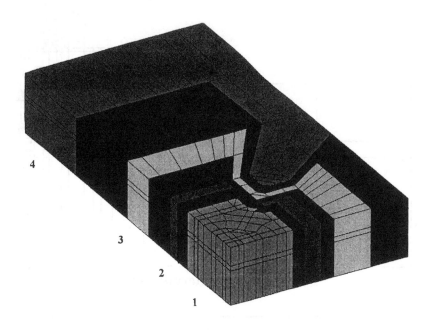

**Figure 3-15. Domain decomposition for foundation
block of Taj Mahal structure (numbers refer
to subdomain numbers)**

The mesh consists of 831 elements (741 brick elements and 90 wedge elements) and 1324 nodes resulting in 3972 global degrees of freedom. The three layers of foundation are modeled using 165 elements (147 brick elements and 18 wedge elements) in each layer. Figures 3-15 and 3-16 show the domain decomposition of the foundation block and the dome using four workstations (four subdomains). Overlapping regions are shown in black. Variation of parallel speedup is shown in Figure 3-17. The CG-EBE-PCG algorithm emerges as a more efficient algorithm for a communication intensive example like the Taj Mahal structure.

3.5 Comments on Performance

A summary of domain decomposition data for example structures (for the case of 4 workstations) is given in Table 3-6. As observed in this table, the domain decomposition algorithm balances the number of nodes. But elements are left unbalanced after domain decomposition. The imbalance, however, is not severe.

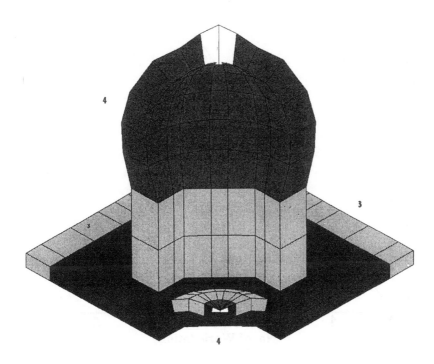

Figure 3-16. Domain decomposition for outer and inner domes of Taj Mahal structure (numbers refer to the subdomain numbers)

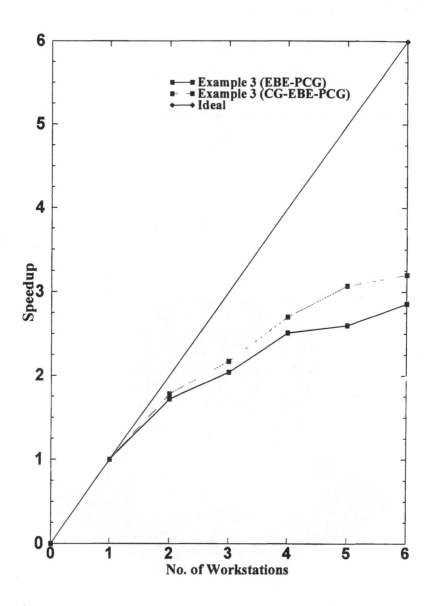

Figure 3-17. Parallel speedup for Example 2

Table 3-7 Summary of connectivity data for the example structures

Connectivity/ Adjacency list	Plane stress problem (Example 1a)	100-Story space structure (Example 2)	Taj Mahal Structure (Example 3)
Average number of elements adjacent to an element	7.8	17.7	16.8
Average number of nodes adjacent to a node	15.4	10.6	19.4
Average number of elements adjacent to a node	2.6	9.6	4.9
Maximum number of elements adjacent to an element	8	20	32
Maximum number of nodes adjacent to a node	21	12	33
Maximum number of elements adjacent to a node	4	11	10

A more significant factor is the percentage overlap which indicates the amount of redundant computations being performed in the PCG iterations. As Table 3-6 shows, this can be as high as 40% for complicated solid models. For skeletal (truss and frame) structures, however, the overlap percentage is small and the amount of redundant computations negligible.

The basic connectivity data is summarized in Table 3-7. The number of nodes involved in a local inter-process communication for Example 3 is given in Table 3-8. In this table, the entry A/B in row i and column j indicates that workstation i sends (receives) the components of the global direction vector associated with A (B) nodes to (from) workstation j. Table 3-8 is symmetric about the diagonal, as expected.

An interesting piece of information generated in the domain decomposition process is the average number of nodes involved in a local communication and number of local communications in a PCG iteration (Table 3-6). Both are very good indicators of the communication requirement of the PCG algorithm for any given example. Since in SPMD model of distributed computing the workstations reach the communication-intensive sections roughly simultaneously, a lower value of the total number of local communications means a lower level of congestion and collision on the network. A lower number for the average size of the communication packet indicates an overall lower volume of communication. The 100-story highrise structure has the best characteristics as far as communication requirement is concerned.

Three factors that affect the efficiency of parallelism in the distributed FE model are load imbalance, redundant computations, and the cost of communication and synchronization. Since a perfect load balance is never possible, there is always some idle time lost during barrier synchronizations. In our algorithms load imbalance arises because we have tried to balance the number of nodes mapped to workstations and this leaves the number of elements unbalanced. Redundant computations have been employed in order to avoid costly communications. However, it is the speed of the network communication which is the biggest limiting factor as far as the performance of distributed algorithms is concerned. Thus, our primary objective has been to minimize the communication cost.

Table 3-8 Number of nodes involved in a local inter-process communication for Example 3 using 4 workstations

Workstation Number	1	2	3	4
1		102 / 167		
2	167 / 102		176 / 178	
3		178 / 176		152 / 152
4			152 / 152	

The three dominant factors which determine how fast two workstations can communicate are: latency, bisection bandwidth, and network topology. Latency is the overhead time required for the system to send and receive a message of zero length. Bisection bandwidth measures the speed with which data can be sent between workstations. A good parallel machine has low latency and high bandwidth. There is always some loss of efficiency due to the collision of packets on the network. Frequency of collision is influenced by the topology of the interconnection network and becomes the dominant factor as the number of workstations grows. For ethernet connected workstations, there is no parallelism in communication due to the bus-like nature of the network topology. If two workstations send certain data packet over the network at the same time, their requests will be serialized leading to varying delays because there can be only one message on the bus at a time.

Table 3-9 Latency and bandwidth of a workstation cluster versus a parallel computer

Machine	Latency	Bandwidth
RS/6000-Ethernet	3.5 milli-sec	50-100 Kbytes/sec
Intel Touchstone DELTA	130 μs	5-7 Mbytes/sec

Table 3-9 shows the comparison of latency and bandwidth of communication between IBM RS/6000 workstations connected by an ethernet network and a distributed memory machine, Intel Touchstone DELTA. It is observed that a workstation cluster connected via ethernet has a latency almost 30 times slower than a massively parallel machine such as DELTA and a bandwidth at least an order of magnitude smaller than that of DELTA. The primary reason for the poor latency of a workstation cluster is the operating system. While RISC architectures have increased the processing power substantially, efforts to speedup the performance of the operating system and network protocols have yet to come to fruition.

However, with rapid developments taking place in networking technology, most of the limitations regarding bandwidth and network interconnection topology are likely to be alleviated soon. Emerging networks such as ATM (Asynchronous Transfer Mode) and Fibre Channel provide scalable bandwidths. For example, Fibre Channel provides a bisection bandwidth up to 50p Mbytes/s for a switched network of p nodes (Duke, 1993). The emerging networking technologies support a vision of a future high-performance computing environment that is far more widely distributed than imagined a few years ago. It may be noted that with the existing networking technology and message-passing libraries, very good performance may not be observed for medium-grained applications such as the solution of a simultaneous system of linear equations. However, at present, generality and portability of algorithms are of more significance than just the speedup values. The PVM software library used in our implementation can easily be extended to include various kinds of networks in future. This frees engineers from dependence on low level data transfer protocols.

4 Distributed Genetic Algorithms for Design Optimization

4.1 Introduction

Optimization of large structures such as highrise buildings and space stations subjected to actual constraints of design codes such as the AISC Allowable Stress Design (ASD) specifications (AISC, 1989) and Load and Resistance Factor design (LRFD) specifications (AISC, 1994) requires an inordinate amount of computer processing time. In recent years, research has been done on the development of efficient parallel optimization algorithms on shared-memory multiprocessors such as Encore Multimax (Adeli and Kamal, 1992a&b, 1993) and Cray Y-MP 8/864 supercomputer (Hsu and Adeli, 1991, Saleh and Adeli, 1994a,b, 1996).

Parallel algorithms for structural optimization reported in the literature have been mostly restricted to shared-memory multiprocessors. However, with rapid advances in microprocessor technology, a recent trend in supercomputing has been towards distributed memory MIMD machines like CM-5 (Thinking Machines, 1991, Adeli, 1992a&b), where a large number of inexpensive microprocessors are harnessed in parallel using fast interconnection networks. An accompanying development has been the increasing popularity of a cluster of workstations as an alternative environment for high-performance computing. In the latter case, a number of workstations are connected together in a local area network (LAN) using a simple network link like ethernet.

The main difference between a dedicated distributed memory machine and a cluster of workstations is in the speed of communication: workstations communicate slowly. However, for applications that show tolerance toward a slow communication link, a cluster of workstations can provide a much better price/performance alternative. The computing needs of the engineering community can be served efficiently by these inexpensive machines. Among the advantages are lower cost of commodity parts (inexpensive microprocessors, memory and communication hardware), smaller development and debugging cost of applications, and feasibility of user and program level fault tolerance (Dongarra et al., 1993). However, for efficient parallelization on such a loosely coupled coarse-grained system, we need to develop new algorithms that exploit parallelism at a coarse task granularity level and require a minimum amount of communication. The frequency of communication is as limiting a factor as its size.

It is primarily in the search for coarse-grained concurrency that we explore biologically inspired genetic algorithms (GA) for structural optimization (Adeli and Cheng, 1993) instead of conventional optimization algorithms like those based on the optimality criteria method (Khot and Berke, 1984). In the latter approach, due to interdependencies in different design iterations, parallelism can be exploited only at the medium-grained subdomain or substructure level (Adeli and Kamal, 1992a&b). Granularity level of parallelization dictates the communication requirement of an algorithm. On architectures with slow communication links, domain-level parallelization results in unacceptably high communication cost involving the shared nodes on the boundary of a domain (Adeli and Kamal, 1990). Consequently, as the number of processors is increased, the communication cost starts dominating the computing time and deteriorates the performance of the algorithm. In contrast, with genetic algorithms, even though the required number of structural analyses is higher, all iterations of analyses at a design step can be done independently and concurrently (Adeli and Cheng, 1994b). This means, a complete structural analysis task can be assigned to one processor and completed without any need for inter-process communication.

In this chapter, we present a scalable, distributed genetic algorithm for optimization of large structures on a network of workstations. Since 95–98% of the total time in a GA-based structural optimization algorithm is spent in the time-consuming fitness function evaluations involving structural analyses, we present a dynamic load balancing procedure for the distribution of the computing load during this phase. The algorithm takes into considerations the uncertainties in the multiuser, multitasking environment of the UNIX workstations and indeterminate nature of fitness function evaluation. We present performance estimates based on the granularity and parallelization efficiency of our model of distributed computing for structural optimization. Details of implementation of this model on a network of IBM RS/6000 workstations and its application to optimization of large space structures are presented in the following chapter.

4.2 GA-based Structural Optimization

In structural optimization, often the aim is to minimize the weight of the structure subjected to various loadings under certain design constraints. The algorithm presented in this chapter is general; but we restrict our discussion to axial-load structures. For a structure consisting of N_E members classified into M groups, we need to find the cross-sectional areas $\mathbf{A}$ = $\{A_1, A_2, A_3, ...,A_M\}$, such that the total weight of the structure

$$W(\mathbf{A}) = \sum_{i=1}^{M} L_i \rho_i A_i = \mathbf{L'\rho A} \tag{4-1}$$

is minimized, subject to the following stress, displacement, and fabricational constraints:

$$\sigma^l \leq \sigma \leq \sigma^u \tag{4-2}$$

$$\mathbf{d}^l \leq \mathbf{d} \leq \mathbf{d}^u \tag{4-3}$$

$$\mathbf{A}^l \leq \mathbf{A} \leq \mathbf{A}^u \tag{4-4}$$

where ρ is a diagonal matrix of weight per unit volume and L_i is the total length of the members in the group i. The terms $\mathbf{A}^u, \mathbf{A}^l, \sigma^u, \sigma^l, \mathbf{d}^u, \mathbf{d}^l$ represent the upper and lower bounds of the vectors of cross-sectional areas $\mathbf{A}$, element stresses σ, and nodal displacements $\mathbf{d}$, respectively.

Since GAs can be used directly only for unconstrained optimization problems, the objective function for structural optimization needs to be formulated using one of the penalty function approaches. Using the quadratic penalty function approach, the corresponding unconstrained optimization problem becomes (Adeli and Cheng, 1993)

$$\text{Min}\phi(A) = \frac{1}{L_f} \sum_{i=1}^{M} \rho_i L_i A_i +$$

$$\alpha \left\{ \sum_{j=1}^{N_E} \left[\left(\left| \frac{\sigma_j}{\sigma_j^a} \right| - 1 \right)^+ \right]^2 + \sum_{k=1}^{N_D} \left[\left(\left| \frac{d_k}{d_k^a} \right| - 1 \right)^+ \right]^2 \right\} \qquad (4\text{-}5)$$

where i refers to the group of members, j refers to the element numbers, k refers to the displacement constraint, α is the penalty function coefficient, L_f is a factor for normalizing the objective function, N_D is the number of displacement constraints, and

$$\left(\left| \frac{\sigma_i}{\sigma_i^a} \right| - 1 \right)^+ = \text{MAX}\left(\left| \frac{\sigma_i}{\sigma_i^a} \right| - 1, 0 \right) \qquad (4\text{-}6)$$

$$\left(\left| \frac{d_i}{d_i^a} \right| - 1 \right)^+ = \text{MAX}\left(\left| \frac{d_i}{d_i^a} \right| - 1, 0 \right) \qquad (4\text{-}7)$$

$$d_i^a = \begin{cases} d_i^l & \text{for } d_i < 0 \\ d_i^u & \text{for } d_i \geq 0 \end{cases} \qquad (4\text{-}8)$$

For elements subjected to axial compression, the allowable stresses depend on the slenderness ratio λ_i which is a function of the length of the member, l_i, its radius of gyration, r_i, and the effective length factor k:

$$\lambda_i = \frac{k l_i}{r_i} \tag{4-9}$$

The allowable tensile (σ_i^u) and compressive (σ_i^l) stresses are used according to the AISC ASD (1989) code:

$$\sigma_i^a = \begin{cases} \sigma_i^l & \text{for} \quad \sigma_i < 0 \\ \sigma_i^u = 0.60 F_y & \text{for} \quad \sigma_i \geq 0 \end{cases} \tag{4-10}$$

$$\sigma_i^l = \begin{cases} \dfrac{\left(1 - \dfrac{\lambda_i^2}{2C_c}\right) F_y}{\dfrac{5}{3} + \dfrac{3\lambda_i}{8C_c} - \dfrac{\lambda_i^3}{8C_c^3}} & \text{for} \quad \lambda_i < C_c \\[20pt] \dfrac{12\pi^2 E}{23\lambda_i^2} & \text{for} \quad \lambda_i \geq C_c \end{cases} \tag{4-11}$$

where E is the modulus of elasticity, F_y is the yield stress of steel, and C_c is the slenderness ratio dividing the elastic and inelastic buckling regions given by

$$C_c = \sqrt{\frac{2\pi^2 E}{F_y}} \qquad (4\text{-}12)$$

Another approach for transforming a constrained optimization problem to an unconstrained optimization problem is the augmented Lagrangian approach (Powell, 1969 and Fletcher, 1975). In this case, instead of a single penalty function coefficient, two parameters associated with each design constraint are used. These parameters are periodically adjusted according to the information obtained from the previous iterations. Adeli and Cheng (1994a) present a genetic algorithm for optimization of structures using the augmented Lagrangian approach. In this case the penalty function is defined in the following form:

$$\text{Minimize } \phi(\mathbf{A}, \gamma, \mu) = \frac{1}{L_f} \sum_{i=1}^{M} \rho_i L_i A_i +$$

$$\frac{1}{2} \left\{ \sum_{j=1}^{N_E} \gamma_j \left[\left(\frac{|\sigma_j|}{|\sigma_j^a|} - 1 + \mu_j \right)^+ \right]^2 \right\} +$$

$$\sum_{k=1}^{N_D} \gamma_{k+N_E} \left[\left(\frac{|d_k|}{|d_k^a|} - 1 + \mu_{k+N_E} \right) \right]^2 \qquad (4\text{-}13)$$

where γs are the Lagrangian multipliers and μs are the constraint parameters associated with stress and deflection constraints.

4.2.1 Objective function evaluation

In order to evaluate the objective function represented by Eq. (4-5) or Eq. (4-13), a structural   analysis needs to be performed. This consists of evaluating element stiffness matrices, their assembling to the global stiffness matrix $\mathbf{K}$, and solution of the resulting system of linear equations

$$\mathbf{Kd} = \mathbf{P} \tag{4-14}$$

where $\mathbf{P}$ is the vector of external nodal loads.

Of these steps, the solution of the system of linear equations, Eq. (4-14), is the most time-consuming part. On smaller computers and workstations, the choice of the equation solver is often decided by its memory requirement since $\mathbf{K}$ is a large and sparse matrix for large structures. In finite element analysis, the frontal solver (Irons, 1970) based on the Gaussian elimination approach is the most widely used method. In recent years, the iterative preconditioned conjugate gradient (PCG) method (Golub and Loan, 1989) has become increasingly popular. In the latter approach, the global stiffness matrix need not be assembled and most of the computations can be done on an element-by-element basis (Winget and Hughes, 1985).

In this approach, the processing time and memory requirements grow linearly with the number of elements (in contrast to direct solvers where they grow polynomially). Thus, even for large problems there is no need for out-of-core data storage (as required in the case of the frontal solver). In a networked computing environment, hard disks of the workstations are often networked into one common logical unit through a file server. While reading from disk can proceed concurrently, writing to the common area of disk space may lead to conflict and, hence, inconsistency. Therefore, in this book we use the PCG method with a diagonal (Jacobi) preconditioner without any need for out-of-core data management for large structures consisting of a few thousands members.

4.2.2 Genetic search procedure

A GA works on a population of q finite length strings called chromosomes. A chromosome most often is just a bit string, a list of 0s and 1s. The length of each string in a population is the same and fixed according to the required accuracy. Chromosomes represent the design alternatives or candidate solutions in some logical way. For the structural optimization problem, the design variables are floating point numbers for cross-sectional areas, which must be encoded as binary digit strings. Using the decoded value of the parameter set as input, the fitness value associated with each string is evaluated. Based on their fitness function values, two chromosomes are selected at a time and genetic crossover and mutation operators are applied resulting in generation of two offspring that form the next generation of the population.

A GA-based structural optimization algorithm requires formulation of a constrained optimization problem as an unconstrained maximization problem, encoding of design parameters (cross-sectional areas) as bit strings, evaluation of fitness of each string in the population (requiring a structural analysis) and population regeneration using genetic operators such as reproduction, crossover, and mutation (Adeli and Cheng, 1993).

4.3 Distributed Genetic Algorithm

The CPU time required for each design iteration consists of two main parts. The first part is fitness function evaluation. Since one structural analysis is required to obtain the fitness value of each string, in each design iteration, there will be as many analyses as the number of strings in the population. Assuming that each fitness function evaluation takes t seconds, the total CPU time for fitness function evaluations at a design iteration is qt seconds. The second part is the CPU time consumed by the genetic operators such as reproduction, crossover, and mutation. We denote it by g. On a single sequential machine, the total CPU time required for s iterations of design optimization is expressed by the simple relation:

$$T_{seq} = (qt + g)s \qquad\qquad (4\text{-}15)$$

While actual values vary according to the problem size, the ratio g/qt is usually very small (less than 1%) and g is negligible compared to qt. To speed up the computations, attention should be focused on reducing the processing time required for fitness function evaluation that depends on the number of strings in the population and time required for each fitness function evaluation. None of them can be decreased below a certain level. A large population size ensures that the search is not confined in a sub-space, thus giving GA its robustness and power to deal with multimodal objective functions. Thus, q cannot be reduced substantially without the risk of getting trapped in a local minimum. Similarly, processing time for fitness function evaluation (t) depends to a great extent on the efficiency of the linear equations solver, which cannot be improved beyond a certain level on a single machine.

Consequently, distributed processing on a network of machines becomes an attractive approach to reducing the total execution of GA-based structural optimization. The distributed genetic algorithm presented in this chapter is based on the distribution of work load among a number of machines during the time-consuming fitness function evaluation phase followed by a single central population regeneration (Figure 4-1). Our model of distributed GA exploits the fact that each string in the population represents an independent coding of the design parameters and hence its fitness function evaluation can be done independently and concurrently.

4.3.1 Dynamic load balancing

The simplest approach to load balancing is partitioning of the load off-line (i.e., before the job is started or at an early stage of its execution). This is called static load balancing and is the approach most often used in parellelizing most applications. Domain decomposition of structural models (Adeli and Kamal, 1993) is an example of static load balancing. For the current problem, we need to distribute q fitness function evaluation tasks on p machines. Theoretically, load balance can be achieved by the distribution of tasks in the ratio of performance ratings of the machines. Assuming all machines are identical, we can theoretically assign q/p tasks to each machine in the beginning.

However, this simple approach to load balancing is satisfactory only if

(a) q is divisible by p,

(b) performance rating of machines is available and accurate,

(c) all tasks take the same amount of time,

(d) machines are dedicated, and

(e) there is no fault in the system (network or machines).

In a networked computing environment, any of these desirable features may be missing. A severe load imbalance will arise when some of the above conditions are violated.

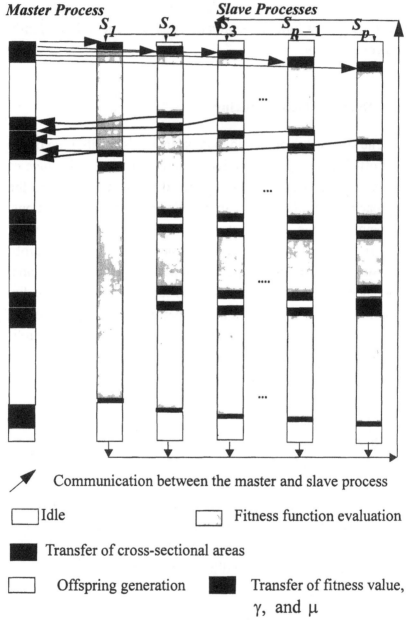

Figure 4-1. Schematic representation of the distributed synchronous GA

In a distributed GA for structural optimization, the following factors lead to a load imbalance:

I) *Heterogeneity in machines.* Very often, a computer laboratory is equipped with machines from different vendors in different models. Even two machines from the same vendor may differ in performance characteristics by an order of magnitude.

II) *Unequal CPU time requirement of tasks.* The major processing time of a single fitness function evaluation is in the solution of the system of linear equations represented by Eq. (4-14). The number of iterations required by the iterative PCG method used in this chapter (because of their low memory requirement) is not known in advance. Convergence of the PCG method depends on the condition number $C(\mathbf{K})$ of the $\mathbf{K}$ matrix defined as $C(K) = \lambda_{max}/\lambda_{min}$, where λ_{max} and λ_{min} are the largest and smallest eigenvalues of K, respectively (Golub and Loan, 1989). $C(\mathbf{K})$ among other things is influenced by the magnitude of cross-sectional areas of the elements themselves. In our problem, for various structural analyses of the structure at a given design iteration, $C(\mathbf{K})$s differ because cross-sectional areas of elements encoded as strings are generated randomly by the genetic search operators. Hence, the number of iterations in the PCG method is unpredictable and its processing time requirement varies accordingly.

III) *Variable processor load in a multiuser, multitasking environment.* Work loads on machines vary with time. On weekends and late nights, labs are under-used and hence, most machines are lightly loaded or idle. At other times, machines may be busy due to multiple tasks being run by several users. In such situations, the CPU will be time-sliced among processes contending for it, leading to variable execution time for identical tasks.

Thus, in all likelihood, static load balancing will result in poor performance for a distributed GA in a networked computing environment. Consequently, we explore the possibility of on-line or dynamic load balancing: the process of making run-time decisions about distribution of work load among processors. The idea is to keep all the processors busy as much as possible between two synchronizations (population replications). More powerful and lightly loaded processors are idle more often and hence, should be assigned more of the work load. Of course, dynamic load balancing may lead to poorer performance if time spent performing load balancing is not offset by an even greater time savings by reducing the variance in execution times of the computation on various processors (Quinn, 1993). We present a dynamic load balancing procedure employing the 'processor farming' concept (Geist et al., 1993).

The distributed genetic algorithm is organized in a master-slave configuration with one master process and a number of slave processes. Figure 4-2 shows the distribution of work load and the communication pattern between the master and slave processes. Process is a unit of execution and refers to a copy of the program in memory, currently executing on a processor or machine. A process performs a number of tasks. In this work, task refers to a fitness function evaluation.

In the processor–farming model, the master process creates and holds the 'pool' of fitness function evaluation tasks and assigns them to the slave processes dynamically, as the slave processes fall idle. Figure 4-3 shows the flow diagram for the master process. Slave processes perform the fitness function evaluations concurrently and report back the values to the master process. Figure 4-4 shows the flow diagram for a slave process. The master and the slave processes cooperate through message-passing.

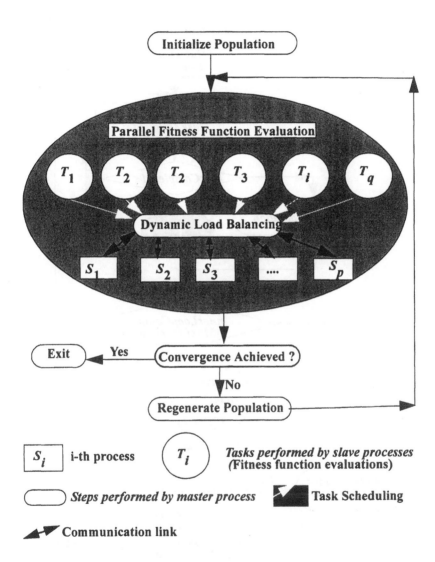

Figure 4-2. Sequence of data transfer and task distribution

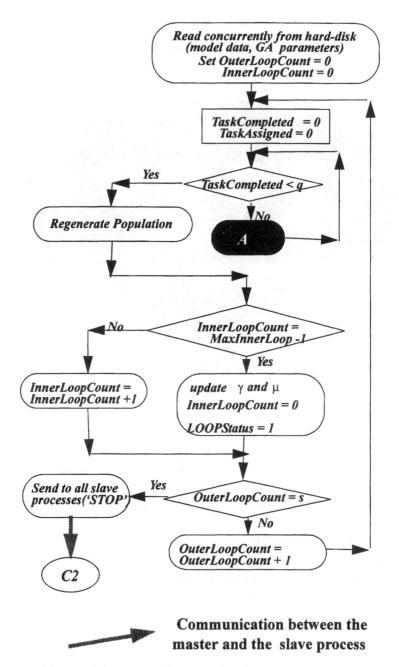

Communication between the
master and the slave process

**Figure 4-3. Flow diagram for the master process in the
distributed augmented Lagrangian GA**

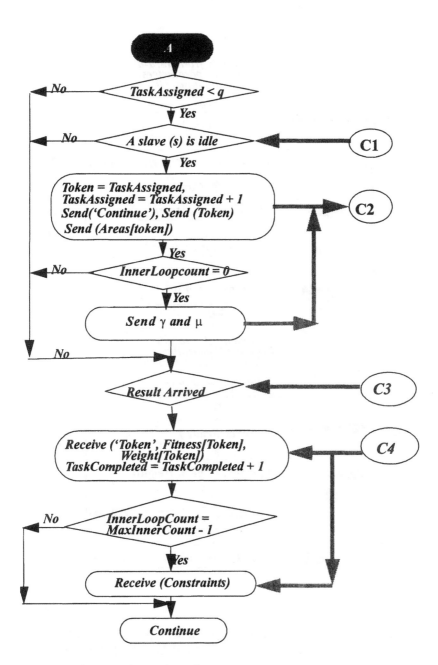

Figure 4-3 continued

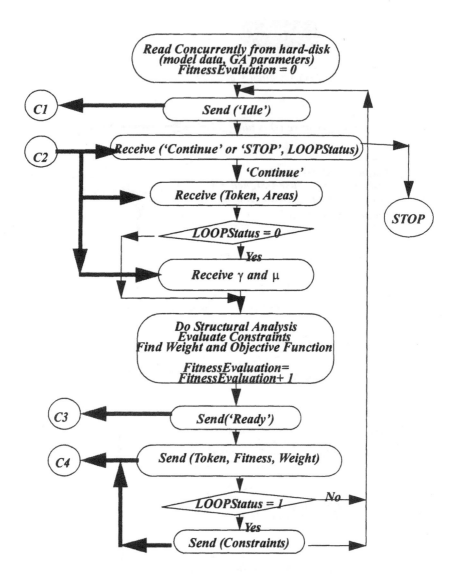

Figure 4-4. Flow diagram for slave processes in the distributed augmented Lagrangian GA

First, a slave process sends a message to the master process (C1 in Figures 4-3 and 4-4) indicating it is idle (in the beginning, all the slave processes are idle) and waits for a message from the master process to continue or stop. If there is a need to evaluate the fitness value for a string in the current population, the master process sends a 'continue' request to the slave process (C2 in Figures 4-3 and 4-4). It also sends two other pieces of information. First is a tag or token number which is used to identify a string while sending the decoded areas and receiving the fitness value. This is done to account for the fact that the order in which slave processes finish fitness function evaluation tasks is not the same as the order in which they are assigned these tasks. Second is the array of decoded cross-sectional areas (A) of the string. The set of cross-sectional areas serves as input to the fitness function evaluation task. The output from a slave process, the fitness value and associated weight of the structure, is sent back to the master process (C4 in the Figures 4-3 and 4-4).

The process of assigning fitness function evaluation tasks and retrieval of fitness values is performed continuously. In other words, the master process goes on in a 'busy wait' loop looking constantly for an idle slave to assign a task to (message tagged C1 in Figures 4-3 and 4-4) or arrival of the fitness value from a slave process (message tagged C3 in Figures 4-3 and 4-4). When all the fitness function evaluations at a given design iteration have been completed, the master process performs the population replication while the slave processes wait. Further, in the augmented Lagrangian GA, the master process updates the parameters γ and μ at the end of the inner loop level.

4.3.2 Quality of search

In developing a distributed GA, a trade-off has to be made between the degree of parallelism and genetic search quality. Synchronous GA with centralized population produces exactly the same sequence of population generations as the sequential GA and, therefore, finds the same solution as the original GA (Pettey et al., 1987). There is no degradation whatsoever in the search quality due to parallelization, but there exists a small part (population replication) which is sequential and acts as a bottleneck to the performance of the distributed algorithm.

A few alternative distributed GAs have been suggested in the literature (Hoffmeister, 1991, Bianchini and Brown, 1993). In all of these, independent processes responsible for handling small subpopulations are used (with periodic communication among the processes). While this approach improves the degree of parallelization, it also degrades the quality of search, since processes may get stuck in local minima. In contrast, in the centralized population model presented in this chapter, the master process has global knowledge about the search process which results in a faster convergence toward the global optimal solution. Consequently, results are obtained in a smaller number of population generations or design iterations. This approach is more suitable for optimization of engineering systems where a fitness function evaluation is often a simulation that usually takes a long and unpredictable amount of time.

4.4 Communication Cost and Performance Estimates

In this section we estimate performance characteristics of our distributed GA for structural optimization. The development is based on the simplifying assumption that all processors are dedicated and homogeneous. First, we present the estimates for the GA based on the penalty function method. Subsequently, we discuss the issues specific to the augmented Lagrangian GA.

4.4.1 Granularity

Task granularity refers to the computations-to-communications ratio. The larger the task granularity, the greater is the degree of parallelization. Communication costs in a single fitness function evaluation in the distributed GA involving the master process and one slave process are due to

a) a message (an integer) from the slave process to the master process indicating that it is idle (C1 in Figures 4-3 and 4-4),

b) a message (an integer) to 'continue' from the master to the slave process (C2 in Figures 4-3 and 4-4),

c) transfer of decoded areas (M floating point numbers) from the master to the slave process (C3 in Figures 4-3 and 4-4), and

d) transfer of the fitness value and the weight of the structure (two floating point numbers) from the slave process to the master process (C4 in Figures 4-3 and 4-4).

Thus, there are a total of four communications between the master and a slave process for each fitness function evaluation. The time required to communicate a data packet (a group of data transformed into a number of bits transmitted collectively on the communication link) between two processors consists of two parts: setup time (also called latency) and transfer time (called bandwidth). For a packet of size n bytes, total communication time is expressed as $t_c = t_s + n/t_w$, where t_s is the setup time and b_w is the data transfer rate on the wire, in bytes per second. For a cluster of workstations connected through an ethernet network, latency is approximately 3.5 milli-sec and the peak data transfer rate is $b_w = 1.2 Mbytes$ /sec. Assuming a floating point number is represented by eight bytes and an integer is represented by four bytes (the standard on most UNIX systems), the overall cost of communication for each fitness function evaluation is

$$t_c = 4t_s + 8M/b_w + 8/b_w \qquad (4\text{-}16)$$

It may be noted that the communication cost grows linearly with the number of groups of members in the structure. However, the major cost of communication is the setup time for communications $(4t_s)$ which is relatively independent of other factors such as the size of the data being transferred.

The ratio of cost of communication to the cost of fitness function evaluation(t), $k_2 = t_c/t$, gives a good measure of task granularity. The smaller this number, the less time is lost in communications and the better the performance of the algorithm. This is usually the case with distributed GA developed in this work. The variation of communication cost t_c is negligible. However, the cost of fitness function evaluation (t) grows linearly with an increase in the size of the structure as indicated by the number of elements. Thus, the ratio $k_2 = t_c/t$ decreases with an increase in the size of the structure. For example, for a space dome structure with 131 elements, 61 nodes, and 14 design groups (Section 5.4.2), the time taken for one fitness function evaluation on an IBM RS/6000 workstation is 0.25 seconds and the communication cost estimated using Eq. (4-16) is 0.056 seconds. In this case, $k_2 = 0.224$ and the communication cost is relatively large (around 22% of the total time). However, for a large structure with 848 elements, 224 nodes, and 67 structural groups (Section 5.4.3), CPU time taken for one fitness function evaluation is around 12 seconds and t_c is estimated at 14 milli-sec. Thus, the ratio $k_2 = 0.0012$ is very small, indicating that the time lost in communication is negligible (around 1%). The important conclusion is that the task granularity and hence, performance of the distributed genetic algorithm presented in this chapter improve fast with an increase in the size of the structure.

4.4.2 Speedup

Speedup is defined as the ratio of time taken to execute a task on p processors to the time taken on a single processor with no communication or synchronization primitives inserted in it. Efficiency is defined as speedup divided by number of processors. Deviation from ideal (linear) speedup (and hence, efficiency) is caused mainly by two factors. First, there is a communication cost associated with every fitness function evaluation t_c (estimated by Eq. (4-16)). Second, population replication is performed sequentially by the master process. The ratio $k_1 = g/qt$ gives a measure of the sequential part of the algorithm. Assuming a worst case scenario in which all communications are sequential (due to the bus-like nature of the network), the total time required for the distributed GA presented in this chapter is estimated as

$$T^p_{par} = \frac{tsq}{p} + gs + qt_c s \qquad (4\text{-}17)$$

and the speedup is given by

$$S_p = \frac{T_{seq}}{T^p_{par}} = \frac{qt + g}{qt/p + g + qt_c} \qquad (4\text{-}18)$$

Upon simplification we obtain the expression for efficiency, η, as

$$\eta = \frac{S_p}{p} = \frac{1 + k_1}{1 + p(k_1 + k_2)} \qquad\qquad (4\text{-}19)$$

Figure 4-5 shows the variation of parallelization efficiency of the distributed GA for structural optimization as a function of the number of processors. Various curves in this figure represent different combinations of factors $K1 = 100k_1$ and $K2 = 100k_2$.

The term k_1 decreases with an increase in the size of the structure because the cost of population replication (g) grows linearly with the number of groups in the structure (M), whereas the time of fitness function evaluation (t) varies with the size of the structure. It should be noted that for large structures the total number of elements grows a lot faster than the number of design groups (various types of members in the structure). As estimated before, the factor k_2 also decreases with an increase in the size of the structure.

As Figure 4-5 shows, efficiency increases drastically with a decrease in either k_1 or k_2. For smaller values of the factors k_1 and k_2 (below 1%), parallelization efficiency of higher than 90% is expected. However, for small structures, both k_1 and k_2 are relatively large, resulting in low efficiency as the number of processors grows. Examples presented in the next chapter illustrate these points further.

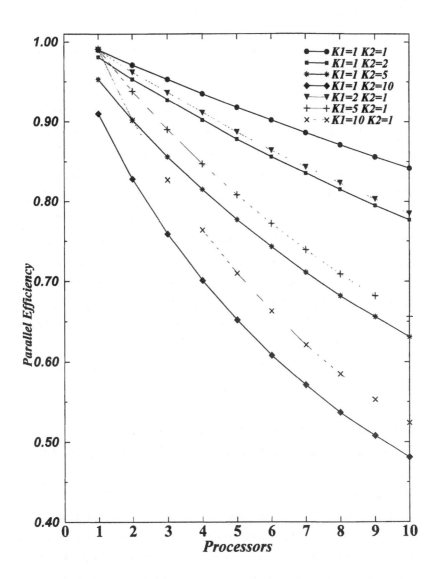

Figure 4-5. Variation of parallelization efficiency with number of processors

4.4.3 Augmented Lagrangian GA

Distributed augmented Lagrangian GA requires some additional considerations. In this case, communication requirements vary with iterations and there is a larger amount of data transfer involved. After every certain number of iterations (MaxInner-Loop in Figure 4-3), the parameters γ and μ are updated by the master process and then redistributed to the slave processes for use in the subsequent iterations. Hence, only the first and the last iterations of the inner optimization loop differ from those in the penalty function-based GA as far as their communication requirements are concerned. To update these parameters the master process needs the values of constraints (computed by the slave processes) for all the strings of the population (their averages are used for updating γ and μ). The number of floating point numbers involved in the data transfer is twice the number of stress and displacement constraints.

For distributing the new values of these parameters (at the beginning of the inner loop), the arrays of γ, μ, and the decoded areas (A) are sent simultaneously to the slave processes while assigning tasks to them (message tagged C2 in Figures 4-3 and 4-4). To transfer the values of constraints from slave processes (in the last iteration of the inner loop), the data is combined with the fitness function value (message tagged C4 in Figures 4-3 and 4-4). Thus, both sending the parameters (γ and μ) and receiving the constraints between the master and the slave processes are incorporated in the existing data transfer mechanism of distributed GA based on the penalty function method. Consequently, the total number of communications is the same in both cases.

Considering that for a small amount of data transfer the setup time dominates the total communication overhead, the additional communication cost in the augmented Lagrangian GA is insignificant. Further, the quality of the search with augmented

Lagrangian approach is much more improved, which more than compensates for slightly higher communication cost (and the resulting lower speedup obtained).

4.5 Final Comments

While a GA-based structural optimization requires a large number of structural analyses, it is parallelizable to a high degree. In this chapter, we presented effective parallelization strategies for GA-based structural optimization using both the penalty function method and augmented Lagrangian approaches. Key to the successful development of algorithms on a cluster of workstations is dynamic load balancing and task scheduling. We presented an effective procedure for dynamic load balancing based on the 'processor-farming' concept which can be implemented in a master-slave configuration. The implementation of the distributed GA optimization algorithms and their application to optimization of large steel structures is presented in the next chapter.

5 Distributed Genetic Algorithms on a Cluster of Workstations

5.1 Introduction

During the past decade there has been an exponential growth in networked computing resources. Local area networks (LANs) have improved in speed by a factor of ten. Microprocessors have doubled in performance every eighteen months and continue to improve in performance at a much greater rate than supercomputers (Turcotte, 1993). Advances in the performance of networks and microprocessors indicate that in the future significant computational capability will be commercially available at a much lower cost. However, the majority of the computing resources included in a network, primarily workstations, may be idle most of the time because of the interactive nature of the applications running on them. As a result, there is growing research interest in making a network of workstations a viable environment for high-performance computing by utilizing their idle CPU cycles. A survey of over sixty software environments for exploiting distributed computing resources is presented by Turcotte (1993).

In Chapter 4 we presented a computational model for distributed genetic algorithms for optimization of large structures. In this chapter we present the implementation of the algorithms on a cluster of workstations connected by an ethernet network. Our algorithms have been developed using PVM (Parallel Virtual Machine), a software library developed at Oak Ridge National Research Laboratory (Geist et al., 1993). A brief overview of PVM and special considerations in the development of distributed genetic algorithms on a network of workstations follows.

5.2 PVM System

PVM 3.2 is a software system that enables a collection of heterogeneous UNIX computers (such as massively parallel machines, vector supercomputers, and scalar machines such as workstations and personal computers) to be used as a single, large concurrent computational resource. Under PVM, a user-defined collection of computers (called hosts) appears as one large distributed-memory computer called a *virtual machine* (Figure 5-1). PVM provides functions for process control and message-passing for communication between processes running on different hosts. Applications written in C, Fortran, and C++ can be parallelized using message-passing constructs.

PVM is composed of two parts. The first part is the so-called daemon process which resides on all the computers making the virtual machine. The daemon is a program that runs in the background and handles requests for message-passing. When a user wants to run a PVM application, he executes daemon on one of the machines; daemons on other machines can be started interactively from the same machine, thus making a user-defined virtual machine. The PVM application can be started from any of the hosts in the virtual machine. The second part of the PVM system is the library of interface routines. Library functions exist for message-passing, creating new processes, coordinating tasks through barriers and formation of groups, and for dynamically modifying the virtual machine itself.

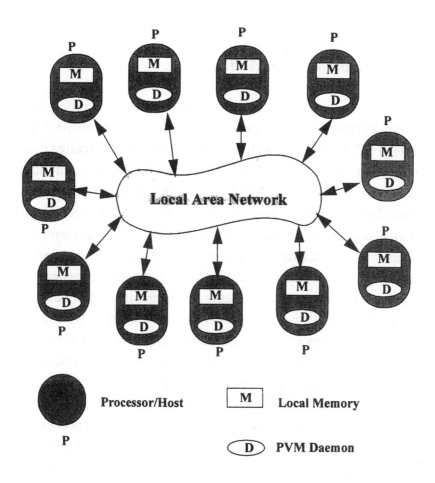

Figure 5-1. Parallel virtual machine

It should be noted that all communications on the ethernet network are sequential. If two processors try to send data packets over the network at the same time, their requests will be serialized, leading to varying delays. This is due to the bus-like nature of the network topology: there can be only one message on the bus at a time.

5.3 Implementation Model

PVM supports two main models of distributed processing: crowd computational model (called Single Program Multiple Data or SPMD model) and tree computational model (called host-node or master-slave model). There is no built-in dynamic load balancing mechanism in PVM. Our model of distributed genetic algorithms for structural optimization is based on the tree computational model and a dynamic load balancing method presented in Chapter 4. The master process performs the population replication and dynamic load balancing while p slave processes perform q fitness function evaluation tasks ($q \gg p$) concurrently at each design iteration. If there are p workstations available, we use one as the master processor for executing the master process. The remaining p-1 machines execute as many slave processes. Since the master processor primarily performs computationally inexpensive integer operations for population replication, its throughput can be improved by executing a copy of one slave process on it. The CPU of the master host is now time-sliced between two processes running on it. Figure 5-2 shows the distribution of master and slave processes along with the daemons on the workstation cluster.

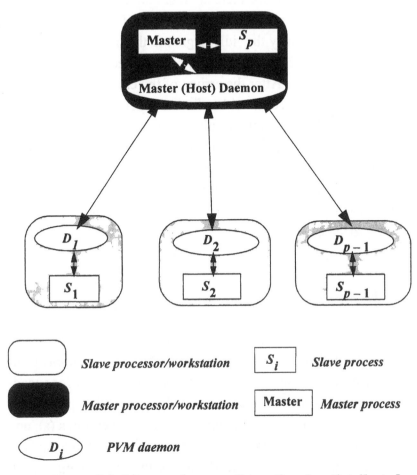

Figure 5-2. Master-slave configuration for distributed GA

PVM provides functions for both synchronous and asynchronous communications. Both blocking and non-blocking receive functions exist. In a non-blocking receive function, the calling process does not wait for the message to arrive. If the message has arrived, the receiving process unpacks it; otherwise, it continues unstopped. In contrast, a blocking receive causes the receiving process to wait and stop until the requested message arrives.

In the distributed GA, there are four communications for each fitness function evaluation (Section 4.3.1). Each slave process sends asynchronously the 'idle' signal to the master process and blocks itself waiting for the arrival of cross-sectional areas from the master process. After the arrival of the cross-sectional areas, a slave process proceeds with a fitness function evaluation, and upon completion sends the results asynchronously to the master process. The master process performs a non-blocking 'busy wait' for arrival of the 'idle' message from slave processes and the results of fitness function evaluation. In the meantime, it sends asynchronously the decoded cross-sectional areas to the 'idle' slave processes.

5.4 Example Structures

Four examples of optimization of structures are presented. In all the examples, a mutation probability of 0.0005 is used. Two-point crossover is applied and the number of bits used for encoding of cross-sectional areas is limited to 16. Other parameters such as population size (q), number of design iterations (s), and bounds for cross-sectional areas vary with each example and are given in Table 5-1. The general data for the four examples such as the number of elements, number of nodes, stress constraints, and the number of design groups are also given in Table 5-1.

Table 5-1 Summary of data for example structures

	Example 1	Example 2	Example 3	Example 4
	17-member truss	Geodesic dome space truss	50-story space structure	35-story space structure
No. of elements	17	132	848	1262
No. of nodes	9	61	224	324
Allowable tensile stress σ^u (MPa)	340.0	170.0	204.0 $(0.60F_y)$	170.0
Allowable compressive stress, σ^l (MPa)	−340.0	−170.0	As per AISC ASD Specifications (Section E2)	−170.0
Unit weight of material (kN/ m^3)	72.75	27.14	77.04	27.14
Young's modulus of elasticity, E (GPa)	207	69	199.95	69

Table 5-1 continued

	Example 1	Example 2	Example 3	Example 4
	17-member truss	Geodesic dome space truss	50-story space structure	35-story space structure
Range of cross-sectional areas (cm^2)	0.645-129	0.645-25.8	0.645-2580	0.645-258
Number of design groups (M)	17	14	67	72
Size of population (q)	100	100	200	140
No. of design iterations (s)	60	60	60	50
Minimum weight obtained in this chapter (Augmented Lagrangian method)	11.54 kN	831.1 N	107.10 MN	1.83 MN
Minimum weight reported previously	11.48 kN Khot and Berke (1984)	820.1 N Adeli and Kamal (1992b)	new example	1.87 MN Adeli and Cheng (1994b)

5.4.1 Example 1: 17-member truss

Example 1 (Figure 5-3) has been studied by Khot and Berke (1984). This is a small structure with 17 members and 9 nodes and has been included here for the purpose of comparison. Design is limited by both stress and displacement constraints. All deflections are limited to 5.08 cm (2 in.). Loading consists of a single vertical downward load of 445 kN (100 kips) at the free end. Members have independent cross-sectional areas, resulting in 17 design variables.

Figure 5-4 shows the variation of the weight associated with the string with the highest fitness value in each generation of augmented Lagrangian GA. The optimal weight obtained after 60 design iterations is 11.54 kN. Khot and Berke (1984) report a minimum weight of 11.48 kN using the optimality criteria method.

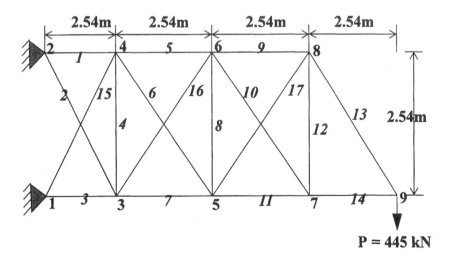

Figure 5-3. Example 1 (17-member truss)

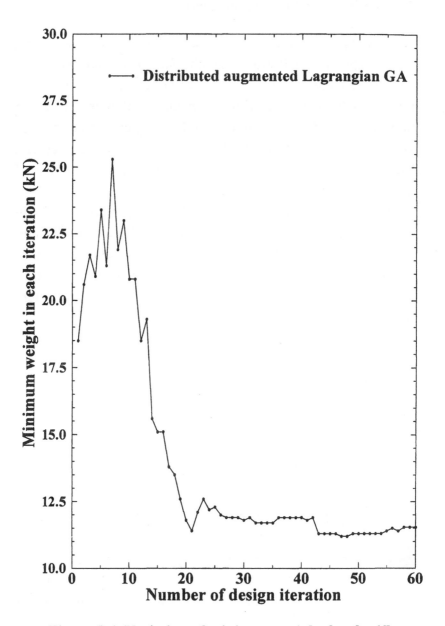

Figure 5-4. Variation of minimum weight for the 17-member truss (Example 1)

Table 5-2 Cross-sectional areas of various members for

Example 1 (in cm^2)

Member number	Distributed augmented Lagrangian GA	Khot and Berke (1984)
1	103.41	102.77
2	0.69	0.65
3	78.60	77.80
4	0.71	0.65
5	54.30	51.99
6	36.87	35.87
7	73.10	76.97
8	0.68	0.65
9	47.1	51.25
10	0.74	0.65
11	26.1	26.16
12	0.65	0.65
13	36.2	36.45
14	26.10	25.80
15	33.24	35.85
16	0.69	0.65
17	34.10	35.99

An important consideration in optimum structural design is violation of design constraints. In GA-based structural optimization, there is no guarantee that the minimum weight found at each design iteration is feasible. Thus, the designs must be checked for constraint violation after every so many iterations. For this structure, the only constraint active in the last design iteration is the displacement at node 9 (Figure 5-3) which equals the upper imposed limit of 5.08 cm. A comparison of the cross-sectional areas obtained in this chapter and those reported in Khot and Berke (1984) is presented in Table 5-2.

5.4.2 Example 2: Geodesic space dome truss

The optimal design of the geodesic dome space truss shown in Figure 5-5 has been studied by a number of researchers including Adeli and Kamal (1992b). This is a medium-size structure with 132 members classified into 14 design groups. The loading on the structure consists of vertical downward loads of 4.45 kN (1 kip) at each node. The displacement limits of ±0.1 in (2.54 mm) are imposed on all nodes in the x-, y-, and z-directions.

The iteration history for 60 design iterations is shown in Figure 5-6. Minimum weight obtained in this chapter is 831.1 N. Adeli and Kamal (1992b) report a minimum weight of 820.1 N.

5.4.3 Example 3: 50-story space structure

The 848-element space truss structure (Figure 5-7) models the exterior envelope structure of a 50-story high-rise building (mega structure). The 848 members are classified into 67 groups resulting in 67 design variables. The displacement constraints are given as 45.72 cm (±18.0 in.) in the y-direction at the top level (equal to 0.0025 times the height of the structure). The loading on the structure consists of horizontal loads acting on the exterior nodes of the space structure at every four floors. The horizontal loads in the y-direction at each node on the sides AB and CD are obtained from the Uniform Building Code (UBC, 1991) wind loading using the equation $p = C_e C_q q_s I$, where p is the design wind pressure, C_e is the combined height, exposure, and gust factor coefficient, C_q is the pressure coefficient, q_s is the wind stagnation pressure and I is the importance factor. The value of C_q for the inward face is 0.8 and for the leeward face is 0.5. It is assumed the structure is located in a region with basic wind speed of 112 Km/h (70 mph), resulting in a wind stagnation pressure of 603.53 N/m^2 (12.6 psf). The importance factor I is assumed to be one.

In this example, the actual provisions of ASD code (AISC, 1989) for design of members in tension and compression have been utilized. For compression members, the buckling constraints are taken into account. Variations of the weight of the string with the highest fitness value (minimum weight of the structure) at each design iteration are shown in Figure 5-8. The minimum weight obtained after 60 design iterations is 107.10 MN.

It is fairly easy to incorporate the actual design constraints in the GA-based structural optimization. In mathematical optimization techniques, objective function and constraints often need to be expressed in terms of the cross-sectional areas of the members as the only design variables. Thus, parameters such as slenderness ratio λ must be approximated by some relationship of the form $\lambda = c_1 A^{c_2}$, where c_1 and c_2 are some constants and A is the area of cross-section of the member. This is necessary to facilitate line search and sensitivity analysis in terms of the cross-sectional areas alone. In contrast, the search process in GA-based structural optimization is random and adaptive and there is no line search. The decoded values of cross-sectional areas of the strings from the current generation of population are used directly to evaluate the constraints in a design iteration.

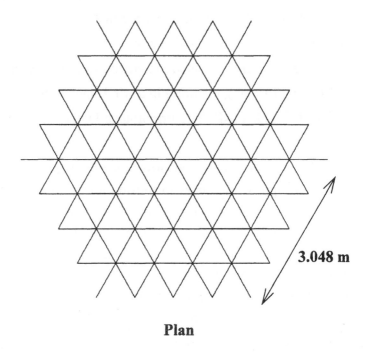

Plan

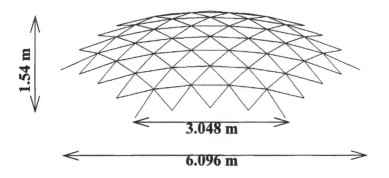

Figure 5-5. Example 2 Geodesic dome space truss

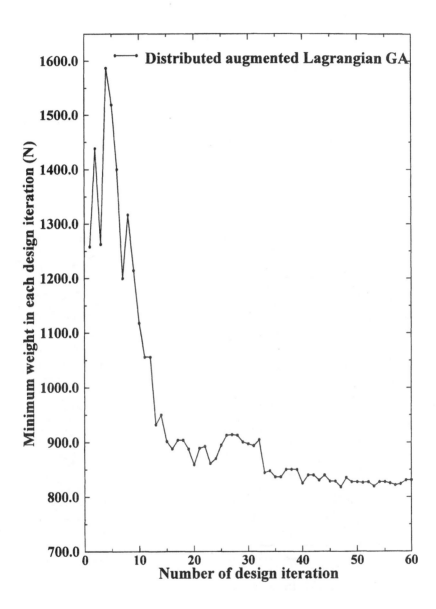

**Figure 5-6. Variation of minimum weight for the
Geodesic dome space truss (Example 2)**

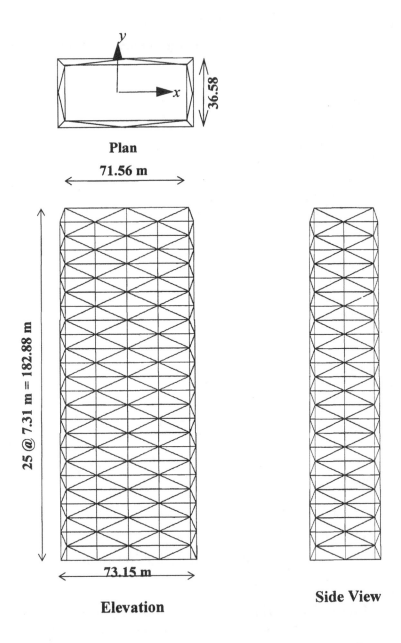

Figure 5-7. Example 3 (50-story space structure)

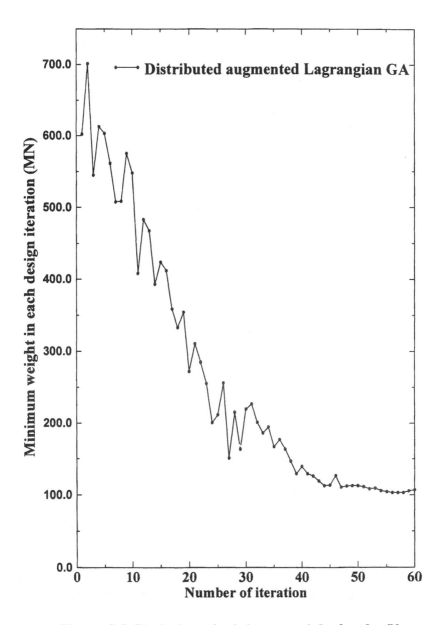

Figure 5-8. Variation of minimum weight for the 50-story space truss structure (Example 3)

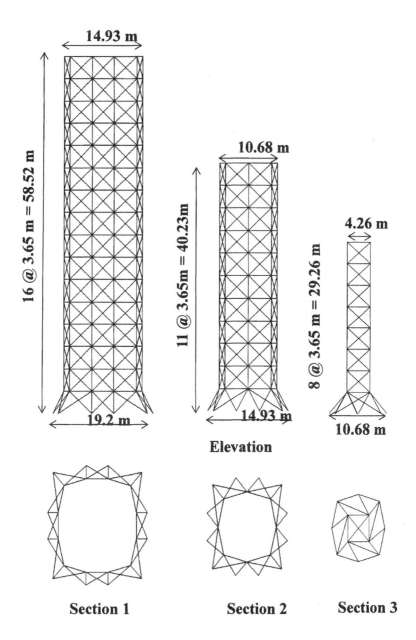

Elevation

Section 1   **Section 2**   **Section 3**

Figure 5-9. Example 4 (35-story space tower)

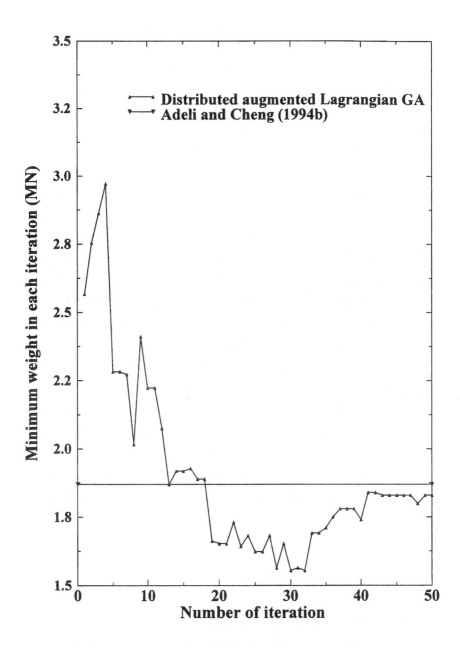

Figure 5-10. Variation of minimum weight for the 35-story space tower structure (Example 4)

5.4.4 Example 4: 35-story tower space structure

This example models the outer envelope of a 35-story space tower structure shown in Figure 5-9. Note that section 2 is located on the top of section 1, and section 3 is located on the top of section 2. This example has been studied in an article by Adeli and Cheng (1994b) using concurrent GAs for optimization of structures on the shared-memory supercomputer CRAY YMP 8/864. This structure consists of 1262 truss members and 324 nodes. Seventy-two design variables are used to represent the cross-sectional areas of seventy-two groups of members in this structure. The loading on the structure consists of downward vertical loads and horizontal loads, details of which can be found in Adeli and Cheng (1994b). The displacement constraints are given as 0.508 m (±20.0 in) at the top of the structure in all three directions (equal to 0.004 times the height of the structure).

The optimal design using the distributed augmented Lagrangian GA resulted in a structure with a weight of 1.83 MN after 50 iterations (Figure 5-10). Adeli and Cheng (1994b) report a minimum weight of 1.87 MN for the same structure using their concurrent GAs. It may be noted that design optimization results using a GA are not exactly reproducible due to the random nature of the search and machine dependency of random number generators.

5.5 Performance Studies

The performance of distributed GAs for optimization of structures is presented in both dedicated (when the lab is not being used by other users) and shared multi-user environments in a laboratory with eleven IBM RS/6000 workstations connected through an ethernet network. The cluster is heterogeneous. Six machines are of Model 320H (numbered 1 through 6) and the remaining five are of Model 220 (numbered 7 through 11). A summary of their performance characteristics is given in Table 5-3. The Model 320H machines are faster than Model 220 machines for the standard benchmark examples. Of the four example structures presented in this chapter, we categorize two (Examples 3 and 4) as large structures for which task-granularity is coarse and high speedups are expected. Example 1 is a small structure and parallelization for this example, in fact, increases the total wall-clock time (also called elapsed time). Example 2 serves as a transition from small to large structures for which interesting observations regarding the performance of distributed algorithms can be made.

First, we compare the performance of our distributed augmented Lagrangian GA on a network of workstations with that of the concurrent GA on the supercomputer CRAY YMP 8/864 reported by Adeli and Cheng (1994b). A comparison of wall-clock times for the two studies for Example 4 (the largest structure) is presented in Table 5-4. It is noted that a cluster of only 11 workstations provides a performance about half of the supercomputer. Part of this performance is attributed to the efficient implementation of the PCG equation solver in our model. Considering that a dozen workstations cost only a fraction of a supercomputer, a cluster of workstations is a more cost-effective alternative to traditional supercomputers for coarse-grained applications such as GA-based structural optimization.

5.5.1 Speedup

All speedup values are measured with respect to a reference machine of Model 220 in the heterogeneous cluster in a dedicated environment. The speedups for Examples 2, 3 and 4 using 2 to 11 workstations are presented in Figure 5-11. It is observed that the speedup of the distributed GA increases with the size of the structure. Maximum speedups of 9.9 and 10.1 are achieved for Examples 3 and 4, respectively, for distributed augmented Lagrangian GA on 11 workstations.

Based on estimates of the computation-to-communication ratio and the sequential part of distributed GA for optimization of large structure, we predicted values in the range of 90% for parallelization efficiency (Section 4.4.2). Observed values of speedups confirm these estimates. For large structures, the algorithms show remarkable scalability in the dedicated environment and there is no sign of degradation of performance with the increase in the number of processors. In spite of poor latency and bandwidth of the ethernet networks, speedups achieved on a cluster of workstation are as high as those observed on dedicated multiprocessors (Adeli and Cheng, 1994b).

The speedup curve for Example 2 reaches asymptotically constant values of 4.6 and 5.0 after seven processors for augmented Lagrangian GA and the penalty function-based GA, respectively. This is attributed to congestion on the network due to the large number of messages bound towards the master processor. Theoretical estimates for speedup of GA-based structural optimization presented in Section 4.4.2 are based on the worst case scenario when all communications are serialized. The predicted value of parallelization efficiency for a medium-size structure is in the range of 50–60% for 7 to 11 hosts, which is close to the observed value.

Table 5-3 Specifications of IBM RISC System/6000

Characteristics	IBM RS/6000 Model 220	IBM RS/6000 Model 320H
Clock Rate	33 MHz	20 MHz
LINPACK[a] (MFLOP)	6.5	9.2
SPECfp92[b] (MFLOP)	34.0	53.1
Memory	39.9 MB (Real) 96.0 MB (Virtual)	32 MB (Real) 128 MB (Virtual)

a. The LINPACK is a simple benchmark of LU decomposition for dense linear systems used to evaluate MFLOP of machines.

b. SPECfp 92 is derived from a set of floating point benchmarks and is used to estimate a machine's peak performance for floating-point operations.

Table 5-4 Comparisons for Example 4

Characteristics	Adeli and Cheng (1994b)	Present study
Population size	140	140
Number of design iterations	50	50
Minimum weight obtained	1.87 MN	1.83 MN
Wall-clock times	4580 seconds (Cray YMP 8/864)	8913 seconds (On 11 workstations)

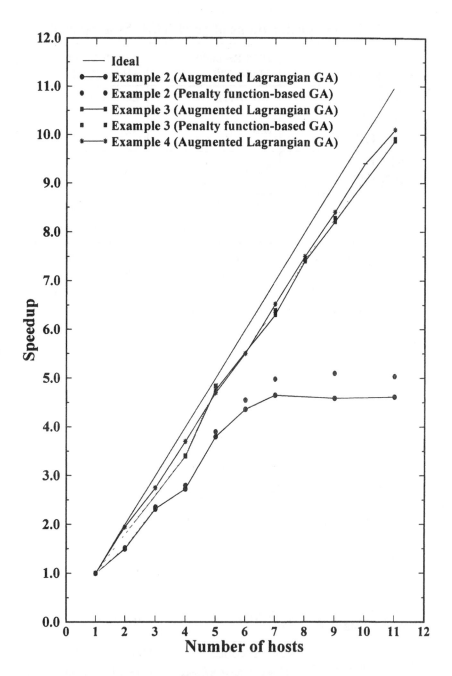

Figure 5-11. Speedups for the example structures in the dedicated environment

A comparison of speedup values observed for the augmented Lagrangian GA with those of penalty function-based GA shows that the former results in slightly lower speedup values (Figure 5-11). As explained in Section 4.4.3 this is due to higher communication costs involved in the augmented Lagrangian GA. However, the difference is insignificant for large structures since the computation-to-communication ratio is very high for such structures and the additional cost of communication in augmented Lagrangian GA is negligible. The difference is more significant for smaller structures (such as Example 2) where communication cost dominates the total time required for GA-based structural optimization.

5.5.2 Processor-farming model

The performance of the dynamic load balancing method based on the processor-farming model is measured by the variation in the number of fitness function evaluations performed by each host. In an ideal case the variation should follow the usage pattern and performance characteristics of the hosts.

The number of fitness function evaluations performed for Examples 2, 3, and 4 by each host in the dedicated environment is shown in Figure 5-12. In a dedicated environment, the distribution of fitness function evaluation tasks (in proportion to the computational power of the hosts) is decided by the performance characteristics of the hosts. Five hosts of model 320H (excluding the master host) numbered 2 to 6, on the average, perform a higher number of fitness function evaluations than the remaining five workstations of model 220, numbered 7 to 11. The master host performs around half as many fitness function evaluations as the slave hosts. This is due to time-slicing or multitasking of its CPU between the master and slave process executing on it. We observe that approximately 45% of the CPU time is dedicated to the slave process on the master host. Thus, the master host performs approximately half as many fitness function evaluations as the slave hosts in addition to performing its other duties as the master host.

Figure 5-13 shows a sample distribution of work load in a shared environment for the three Examples 2, 3, and 4. In this case, a larger distribution of tasks is automatically shifted to lightly loaded machines (for example, host number 6 in Figure 5-13) leaving busy machines (such as number 11 in Figure 5-13) to perform tasks assigned by other users. The total wall-clock times required (using 11 workstations) for Examples 3 and 4 are approximately 40–50% higher in the shared environment compared to the dedicated environment. Actual values vary according to the computational intensity of applications being run by other users.

The variation of load distribution for Example 2 (the geodesic space dome structure, the medium-size structure) shows some interesting results. In this case, the master host performs almost as many fitness function evaluations as the slave hosts. For relatively small structures, the communication cost is high and affects the load distribution significantly. The slave process with a lower communication overhead performs relatively a higher number of fitness function evaluations. This is true for the slave process executing on the master host. Thus, for smaller structures the master host performs a relatively larger proportion of fitness function evaluations compared to the case of large structure (such as Examples 3 and 4).

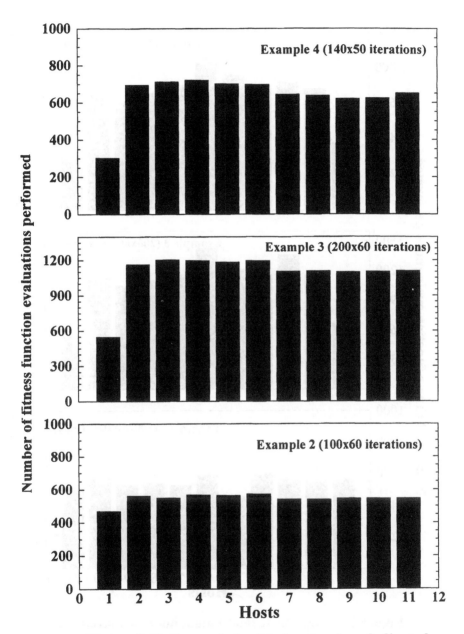

Figure 5-12. Dynamic load balancing in a dedicated environment

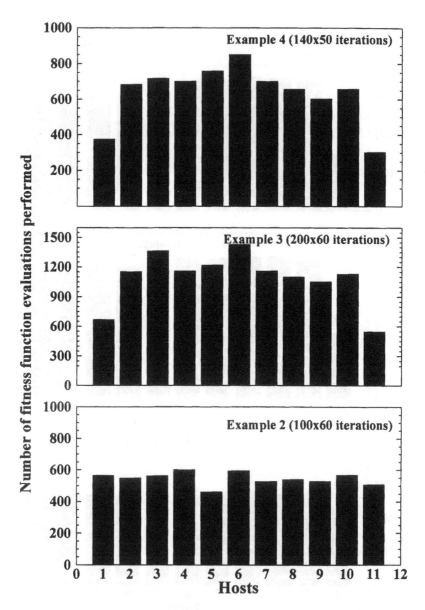

Figure 5-13. Dynamic load balancing in a shared multi-user environment

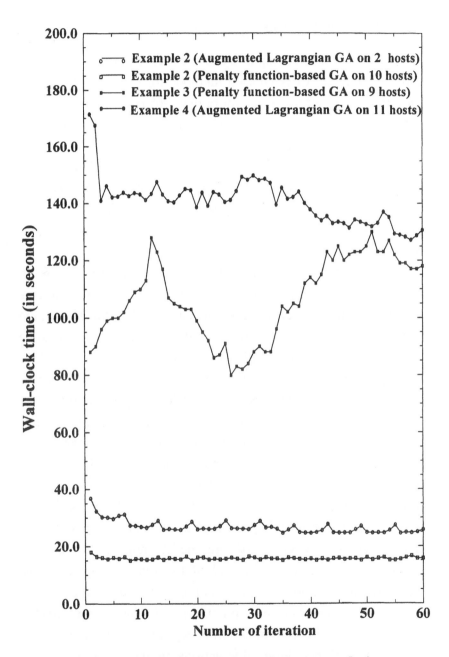

Figure 5-14. Variation of elapsed times per design
iteration

5.5.3 Observations regarding the PCG solver

Performance of the equation solver is critical to the total CPU time requirement of GA-based methods for structural optimization of large structures since more than 90% of the time is spent for solution of the resulting system of linear equations. Figure 5-14 shows the variation of wall-clock times for various design iterations for Example 2 using the augmented Lagrangian GA (for the case of 2 hosts) and the penalty function-based GA (for 10 hosts), Example 3 using penalty function-based GA (for the case of 9 hosts) and Example 4 using augmented Lagrangian GA (for the case of 11 hosts).

Due to the indeterminate nature of the iterative PCG solver, the variation in time requirements of different design iterations is random. Variation is also caused by load imbalance arising in the final stages of a design iteration; there are fewer strings left to be evaluated than number of idle hosts. Figure 5-14 shows that, in general, more fluctuations are observed in wall-clock times of each design iteration for large structures (Examples 3 and 4). Further, due to a well-directed search, fluctuations are smaller in augmented Lagrangian GA compared to penalty function-based GA.

It may be noted that for the augmented Lagrangian GA, the elapsed times required for the first and last iterations of the inner design loop are higher than those of the intermediate design iterations. The reason for this is the higher communication and computation costs involved in updating the Lagrange multipliers and constraint parameters at the end of the inner design loop, as explained in Section 4.4.3. This is more obvious in the case of Example 2 which has a small granularity. For large structures the additional cost of communication is negligible and is masked by the aforementioned large fluctuations in the elapsed time of design iterations. The high scalability of distributed genetic algorithms developed in this chapter demonstrates that a cluster of workstations provides a cost-effective alternative for high-performance computing for coarse-grained applications such as the GA-based structural optimization.

For the structural optimization problem, the tolerance limit for convergence of the iterative PCG solver may be kept relatively high without much loss of accuracy. We found a tolerance limit of 1.0×10^{-5} for convergence of the PCG solver to be fairly accurate for all the examples presented in this chapter.

5.6 Final Comments

In this chapter we presented an efficient implementation of the computational model for distributed GA-based structural optimization using the software library PVM on a heterogeneous cluster of eleven IBM RS/6000 workstations. The distributed GAs were applied to optimization of four structures of various size. GA-based structural optimization is found particularly effective for optimization of large structures based on the provisions of AISC ASD code where design constraints are implicit functions of design variables. We find that the PCG iterative solver is particularly effective for distributed GA on a cluster of workstations due to its low memory requirement if an effective dynamic load balancing method is used to account for its indeterminacy.

Performance of the algorithms was evaluated in both a dedicated and shared multi-user environment of a computer laboratory. For large structures, a high speedup of around 10 was obtained using eleven workstations (meaning a parallelization efficiency of about 90%). Observed speedup of the GAs increases with the size of the model making it particularly suitable for optimization of large structures. The algorithms show no signs of degradation in speedup as the number of processors is increased.

6 Concurrent Structural Optimization on a Massively Parallel Supercomputer

6.1 Introduction

With rapid advances in microprocessor technology, a recent trend in supercomputing has been toward distributed memory massively parallel machines such as Connection Machine CM-5 (Thinking Machines, 1992a; Adeli, 1992a&b), where a large number of inexpensive microprocessors are harnessed in parallel using fast interconnection networks. In this chapter we present a mixed computational model for concurrent computing in structural optimization exploiting both the control parallelism and data parallelism available on the most advanced Connection Machine CM-5 supercomputer.

In Chapters 4 and 5 we presented distributed genetic optimization algorithms for structural optimization on a network of workstations. In spite of the slow speed of communication on ethernet networks connecting the workstations, a high degree of parallel efficiency was achieved due to inherent coarse-grained parallelism of genetic algorithms. In genetic algorithms, at each

147

design optimization iteration, the objective function evaluation involves a large number of finite element (FE) analyses of the structure (one for each string in the population) (Adeli and Cheng, 1993) which can be performed independently and concurrently. This means a complete finite element analysis can be assigned to a processor and completed without any need for inter-process communication. This kind of coarse-grained parallelism can be exploited using the Multiple Instruction Multiple Data (MIMD) model of computing. In this chapter, the MIMD model is organized in the host-node configuration. Each node performs a separate fitness function evaluation involving an FE analysis of the structure. The host process manages dynamically the pool of fitness function evaluation tasks and performs population replication. Message-passing constructs from the CMMD library (Thinking Machines, 1993a&b) are used for communications between the host and node processes using virtual channels.

There is yet another level of parallelism that can be exploited if individual processors are equipped with advanced arithmetic units. Using a data parallel element-by-element preconditioned conjugate gradient (PCG) solution procedure based on the Single Instruction Multiple Data (SIMD) model of computing, the efficiency of the FE analysis can be increased substantially by exploiting the inherent fine-grained parallelism at the element and degree of freedom level. There have been a number of attempts to develop massively parallel algorithms for data parallel finite element analysis on the earlier Connection Machine CM-2. Johnsson and Mathur (1990) developed data parallel iterative algorithms for the solution of the resulting system of linear equations. Farhat et al. (1989) and Belytchko et al. (1990) present algorithms for dynamic and transient simulations while Johan et al. (1992) present algorithms for large scale computational fluid dynamics computations.

In this chapter we present a computational model to exploit parallelism at both coarse-grained and fine-grained levels simultaneously. We employ MIMD control at the outer loop level of design optimization iteration in genetic search and SIMD flow for the solution of the system of linear equations arising in the fitness function evaluation. A data parallel PCG solver has been developed and implemented in CM Fortran (Thinking Machines, 1992b and 1993c) using an effective communication routine from the Connection Machine Scientific Software Library (CMSSL) (Thinking Machines, 1993d) in the scatter-and-gather phase of the PCG iterations. The computational model and resulting algorithm have been applied to optimization of large steel structures.

6.2 GA-Based Structural Optimization

The major cost of GA-based structural optimization (described in more detail in Section 4.2) consists of fitness function evaluation which involves a finite element analysis of the structure for each string in the population. This step consists of evaluating element stiffness matrices (k_e), applying boundary conditions, and the solution of the resulting system of linear equations

$$\mathbf{Kd} = f \tag{6-1}$$

where K is the assembled global stiffness matrix, and f is the vector of nodal loads. Our concurrent GA-based structural optimization algorithm is based on the preconditioned conjugate gradient (PCG) method for solution of the resulting system of linear equations. To exploit the high degree of parallelism a simple diagonal preconditioner has been used instead of more efficient preconditioners based on incomplete Cholesky factorization. The latter method requires factorization of the assembled system of equations which creates a bottleneck on a massively parallel fine-grained machine such as CM-5. In the case of an element-by-element preconditioner of the type suggested by Winget and Hughes (1985), the two-pass matrix-vector multiplication required is inherently sequential. Relative inefficiency of diagonal preconditioner is more than offset by its high parallelizibility at the degree of freedom level.

6.3 CM-5 Architecture and Computing Models

6.3.1 CM-5 architecture

A CM-5 machine consists of tens, hundreds, or thousands of parallel processing nodes, each having its own local memory (currently 32 Mbytes) (Figure 6-1). To improve system utilization, the nodes are grouped into *partitions* of size 2^n, for example, 32, 64, 128, etc. The group of processing nodes in a partition is supervised by a control processor called the *partition manager* (PM) which functions similarly to most workstations. Partition managers run an enhanced version of the Unix operating system called CMOST which provides standard I/O and communication facilities, X Window System, a specialized scientific library called Connection Machine Scientific Software Library (CMSSL), and an integrated programming environment for parallel debugging and performance analysis (called Prism). Other control processors are dedicated to servicing I/O requests and diagnosing component failure.

A CM-5 processing node may optionally contain an arithmetic accelerator consisting of four vector units (placed on two chips) that operate in parallel. Each vector unit controls the associated memory bank of 8 Mbytes through a 72-bit wide data path (Figure 6-2). With a peak memory bandwidth of 128 Mbytes/sec and peak 64-bit floating-point performance of 32 Mflops, the four vector units together provide a memory bandwidth of 512 Mbytes/sec and peak floating-point performance of 128 Mflops. Thus, each CM-5 node itself is a powerful parallel processor. PMs do not have an arithmetic accelerator unit.

CM-5 nodes and partition managers are connected through three scalable high-bandwidth interprocess communication networks called Data Network, Control Network, and Diagnostic Network (Figure 6-1). Access to these networks is provided through a Network Interface (NI) chip. The Data Network is used to transfer large amounts of data at high bandwidth between two nodes. Control Network is used in operations that require synchronization across all nodes, such as reduction (where many data elements are combined to produce a smaller number of elements) and broadcasting (where one node sends a value to all other nodes in the partition). The Diagnostic Network which is hidden from the users is used for system diagnostic and maintenance. CM-5 networks are in the form of a 4-ary fat tree where each node has four children and two parents (Figure 6-3). A fat tree network is advantageous over other types of networks such as hypercube. The tree network can easily be partitioned so that each group of processors has its own portion of the network and the networks do not have to interfere across the partitions. Another advantage is that the failure of a single communication network node does not result in the breakdown of the total network.

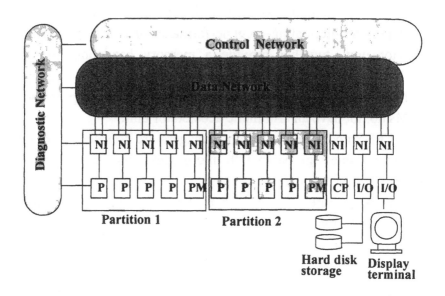

P   Processing Node
CP   Control Processor
PM Partition Manager
NI   Network Interface Chip
I/O Input/Output server

**Figure 6-1. Architecture of the Connection Machine
CM-5**

6.3.2 Programming models

CM-5 supports the two widely used models of parallel programming: the data parallel model which is efficient for synchronization and communication but inefficient for branching, and the control parallel model which is efficient for branching but inefficient for synchronization and communication. Connection machines are designed to operate on large amounts of data which can be highly interconnected such as those found in a finite element analysis or totally autonomous such as those found in database processing. The Connection Machine (CM) architecture can be exploited best by using one of the data parallel programming languages such as CM Fortran, C* or *Lisp (Thinking Machines, 1992a).

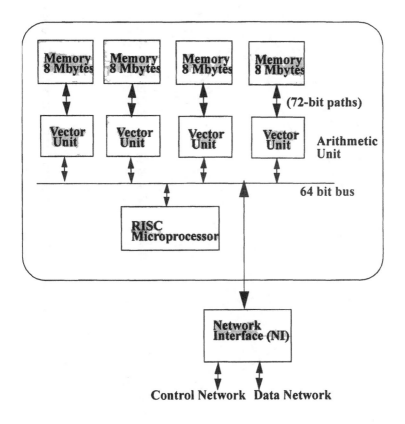

Figure 6-2. A processing node (P) with vector units

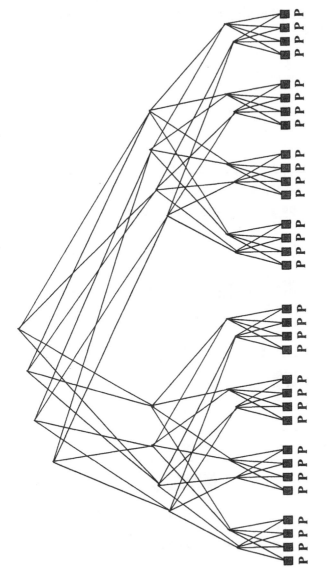

Figure 6-3. CM-5 data network with 32 nodes

In data parallel programming, emphasis is on the use of large and uniform data sets such as arrays whose elements can be processed simultaneously. For example, if A, B and C are arrays containing thousands of elements, a data parallel operation such as $A = B + C$ indicates the simultaneous addition of thousands of elements. Data parallel operations are performed under a single thread of control (called SIMD control). Actual implementation of this feature is hidden from the application developers. They just need to use one of the CM parallel languages to develop data parallel applications. The compilers generate appropriate code to spread an array across the distributed memory of the nodes that operate only on the array elements mapped to them. Parallel variables identical in array dimensions and sizes are laid out in a similar fashion. Thus, corresponding elements of arrays share the local memory of a particular processor and simple operations such as $A = B + C$ require no interprocess communication.

For interconnected data a parallel operation may require movement of data from the local memory of one processor to another. Computations on interconnected data usually proceed in a familiar pattern: first, data elements are brought together (called *gather*) in the local memory of a processor from local memories of other processors, then local computations are performed concurrently. Finally, the result is spread back to the desired memory location (called *scatter*). CM languages have a rich set of operators that can be used for data movement and concurrent computations on interconnected data.

While data parallel programming offers a very versatile and efficient way of harnessing the power of a massively parallel supercomputer like CM-5, most problems in the engineering domain also require good branching facilities such as IF-THEN-ELSE type statements which cannot be parallelized fully using the abovementioned concepts. Effective parallelization of such applications requires control parallelism through multiple threads of control, called MIMD control. In a MIMD control, multiple copies of the program are run independently on the processors. Nodes cooperate through communications. The Connection Machine communication library, CMMD, is an advanced user-extensible communication library that provides traditional message-passing constructs typical of most distributed memory machines.

6.3.3 Execution models

The CM-5 architecture provides for three different models of execution. They are the global model, local (node-level) model, and global/local model (proposed to be implemented in the near future). In the global model the data is laid out across the processors (or vector units). The control processor broadcasts the parallel code blocks to the nodes that operate on the parallel data concurrently. This model requires the use of one of the CM data parallel languages.

In the node-level or local execution model, multiple copies of the program execute independently on different nodes and hence, the data remains local to the nodes (and laid out across the four vector units of the node). The data parallelism can still be exploited at the node-level using one of the CM data parallel languages. The node-level model can be further classified into hostless and host/node models. In the former there is no user-defined process on the host. The host process on the PM runs a server which can be customized to support specific functions such as I/O. This model is ideal for a single program multiple data (SPMD) model of distributed computing. In the host/node model the partition manager runs a user-supplied program on the host.

Through a combination of data parallel programming and various execution models, different models of distributed computing can be implemented. The choice of an execution model is decided by the nature of application, its communication requirement, and the size of the data sets. The global model is suitable when very large data sets are involved. For example, on the smallest 32-node partition with 128 vector units, the size of the array should be around 128*128 = 16384 elements to take full advantage of vector units in the global model of execution. Further, if the data sets are small the memory configuration of the system may not be fully utilized. The portions of the problem which cannot be expressed entirely in the data parallel construct are executed on the PM. Due to the large difference in the performance level of PM and the nodes equipped with vector units, the sequential code blocks executed on the PM may create a major bottleneck. Thus, the global model of programming is ideal only for single large simulations with large data sets and very little sequential code.

Our concurrent GA-based structural optimization algorithm requires a large number of finite element simulations on structures with a few thousand elements and nodes. Based on our estimates of the size of data set involved in a typical FE analysis of these structures, the data set is not large enough to take full advantage of data parallelism using the vector units in the global model of execution. Thus, the host/node model of local (node-level) execution has been selected in this chapter. The local memory limit of 32 Mbyte is sufficient to accommodate the finite element models of highrise building structures studied in this chapter. Further, the data set can be distributed evenly on the four local vector units thus speeding up the computations. Details of parallelization of the distributed GA using MIMD model are presented in the next section followed by a discussion of data parallel fitness function evaluation on the nodes.

6.4 Coarse-Grained Concurrent Genetic Algorithm

Our parallelization of a GA is based on the fundamental premise that each string in the population represents an independent coding of the design parameters and hence, its fitness function evaluation can be done independently and concurrently. The concurrent synchronous GA for structural optimization is based on distribution of work load among processors during the fitness function evaluation phase followed by a single central population regeneration (Section 4.3). This is ideal for structural optimization problems with long fitness function evaluations. Each fitness function evaluation task involves a FE analysis of the structure based on the cross-sectional areas encoded as a string. In each design iteration, the q strings of a population are distributed among p processors. This type of parallelism is referred to as coarse-grained or control parallelism.

The concurrent genetic algorithm is organized in the host/ node configuration (Figure 6-4). The host maintains the centralized population and is responsible for performing genetic search operations such as selection, crossover and mutation, sequentially. Nodes perform fitness function evaluations concurrently. The host manages dynamically the pool of fitness function evaluation tasks and assigns them to nodes as they fall idle. Thus, a dynamic load balance is maintained among processors (Section 4.3). The process of assignment of fitness function evaluation task and retrieval of fitness values is performed continuously. In other words, the host process goes on in a 'busy wait' loop looking constantly for an idle slave to assign a task to or the arrival of the fitness value from a slave process. When all the fitness function evaluations at a given design iteration have been completed, the host process performs the population replication while node processes wait. Further, in the concurrent augmented

Lagrangian GA, the host process updates the parameters γ and μ (Figure 4-3) at the end of the inner loop level.

6.4.1 Virtual channels

The cost of communication between the host and nodes consists of two parts: setup time (also called latency) and transfer time (called bandwidth). For small data transfers on the CM-5 network the setup time required in the CMMD library routines (such as *CMMD_send_block*) dominates the cost of communication. CMMD also provides a much faster means of communication through low-latency unidirectional connection between two processors called *virtual channel*. Channels can be opened between any two processors (once only) and used repeatedly to efficiently transfer the data between the two nodes as long as the pattern of communication remains the same. With repeated use of this low-latency communication channel, the cost of opening and closing the channel is amortized.

Since the pattern of communication between the host and the nodes is known in advance in the concurrent GA-based structural optimization algorithm, virtual channels are used in order to maximize the global communication performance. Each virtual channel has its own descriptor or identifier. Since there are six different communication patterns between the host and any node (Figures 4-3 and 4-4), a total of 6p channels (6 for each node) need to be opened and consequently, as many descriptors are saved in an array on the host.

6.4.2 Number of processors versus the size of population

The size of the population in genetic algorithms is always even. For optimization of structures consisting of a few thousand members classified into a few hundred different types of members, the population size is in the hundreds. The performance of the dynamic load balancing method, as outlined above, depends to a major extent on the relative predominance of the size of the population over the size of partition. If $q \gg p$, dynamic load balancing performs well and idle (waiting) time is small and negligible. However, when q is very close to p, the performance of the dynamic load balancing algorithm degrades and static load balancing may be found more suitable. In the unlikely case when $q < p$, the host-node model becomes inefficient as some of the nodes always remain idle (at any stage, only q nodes are busy and the remaining p-q nodes are idle). In this case, the global mode of execution may be more suitable than the host/node local model. However, for GA-based structural optimization, such a situation can be avoided by choosing the size of the CM-5 partition and population appropriately.

The size of the population is initially selected approximately. A good rule of thumb is four to six times the number of design variables. Next, the partition size is chosen according to the size of population. The partition size should be smaller than the population size (roughly 1/8 to 1/2 times the size of the population). Using the largest available number of processors is often not in the best interest of all the users and often results in reduced overall efficiency (system throughput). For large structures, the size of the population is large and hence, a larger partition size may be used. For a small structure which requires a smaller population size, a 32-node partition may just be sufficient. Finally, if possible, the population size should be rounded up to the nearest higher multiple of partition size to improve overall throughput and quality of search. With these considerations, the dynamic load balancing is found to be an efficient strategy for problems considered in this chapter.

6.4.3 Parallel I/O

Flexible I/O is required to reduce the time spent in reading from and writing to the hard disk, which often takes a substantial amount of time for large structural optimization applications. The message-passing library CMMD offers flexibility through extension of standard Unix I/O utility functions. It provides routines for parallel input/output at the node-level. CMMD provides four I/O modes: local independent, global independent, global synchronous broadcast, and global synchronous independent. The first two allow node-level I/O, with each node acting independently of others. In the last two modes all nodes read and write data in some cooperative fashion which speeds up data transfer substantially.

In our concurrent GA-based structural optimization all nodes read the same input data from the disk at the beginning. Input data include structural data such as information for member connectivity, nodal coordinates, external loading, member grouping, and boundary conditions and genetic parameters such as size of population, number of iterations in outer and inner loops of the augmented Lagrangian genetic algorithm, scaling factor, and probabilities of mutation and crossover. In comparison to direct solvers where there is usually some disk I/O involved, due to the low memory requirement of the iterative PCG solver used in this chapter, there is no disk I/O in this stage of the solution.

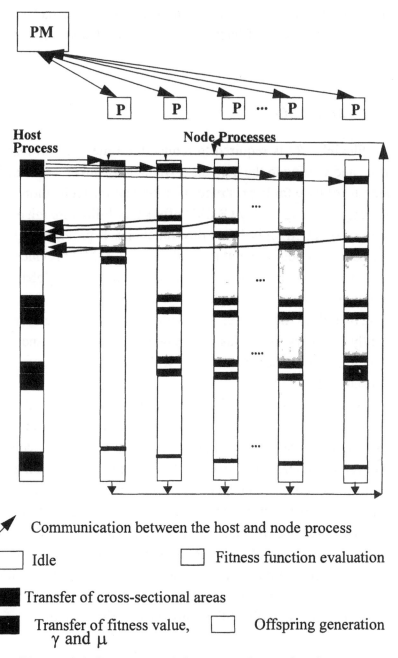

Figure 6-4. Sequence of data transfer and task distribution in the concurrent GA

Since all nodes read the same data from a common file, the global synchronous broadcast I/O mode is used to enhance the I/O performance. In this mode, one node (node 0) reads the data from the file and broadcasts them to all the other nodes in the partition. Since only one node reads the data from the disk, the overhead of disk I/O is reduced substantially.

Parallel I/O operations can be performed by linking the host program to the standard CM file server routine. I/O calls on the nodes result in messages being sent to the I/O server on the host which is scheduled to handle these requests. A typical I/O operation involves a call to the function *CMMD_service_loop* on the host. This starts an I/O server (Figure 6-1) that remains in effect until all nodes have called the function *CMMD_global_ suspend_servers*. All nodes perform the required I/O operations in this time interval (between these two calls) and are synchronized at the end of the I/O operations.

So far we have described the parallelization of the outer loop level of GA-based structural optimization. On CM-5 machines equipped with optional vector units on the nodes, a fast fitness function evaluation routine can be developed using one of the CM data parallel languages to exploit a higher level of parallelism. The development of such an algorithm using the CM Fortran language is described next.

6.5 Fine-Grained Data Parallel Fitness Function Evaluation

The major part of fitness function evaluation consists of the solution of Eq. (6-1). As discussed in Section 6.2, the iterative element-by-element (EBE) preconditioned conjugate gradient (PCG) method is used in this chapter. In this method, there exists a very high degree of concurrency at the fine-grained element and node level. The main computation in a PCG iteration is the matrix-vector product $h = Kp$, where p is the global direction vector for the conjugate gradient search. This step can be performed entirely at the element level without any need to assemble the global stiffness matrix (Winget and Hughes, 1985). Note that in the following discussions the term node is used in two different contexts: structural node (a node in the discretized model of the structure) and processing node of CM-5.

The element-by-element matrix-vector product operation can be parallelized at the element level by mapping an unassembled element to a processing node or a vector unit (Johnsson and Mathur, 1990). The other operations in the PCG iterations involve dot products of the form $v \bullet u$ and vector updates of the form $v \pm ku$ which can be parallelized at the structural node and/ or degree of freedom level by mapping an assembled nodal point to a processing node or a vector unit. There are only four processing elements (vector units) at each processing node and thus the CM run-time system (CM-RTS) (a component of CMOST) maps many elements and nodes of the structure to a processing element. However, from the application developer's point of view the data parallel operations are programmed assuming there are as many processing elements as there are nodes or elements in the structure. This programming model assumes that all the element or node level operations proceed concurrently even though actual processing at the element and node level is distributed and time-sliced evenly across processing elements.

For an element-level matrix-vector product, first the components of p corresponding to the nodes in the connectivity of the element are accumulated into a vector p_e from the assembled nodal points to the set of unassembled finite elements (*gather*) (Figure 6-5). Since the nodes of an element may be mapped to a processing element other than the processing element to which the element itself is mapped, irregular communication is required. Next, a local dense matrix-vector product is performed between the element stiffness matrix (k_e) and the direction vector (p_e) and the result is stored in the local array $h_e = k_e p_e$. Finally, the result is stored back to the assembled nodal points (*scatter*). We briefly describe the main data structure involved in our implementation of the element-by-element iterative solution algorithm.

The most important aspect of the data parallel iterative solution procedure is the data layout. Once various data sets involved in PCG iterations are laid across the processing elements, computations involving them can be performed concurrently using data parallel constructs. The SERIAL directive in the CM Fortran language ensures that all components of an array in a given dimension reside in the local memory of the same processing element or vector unit. The NEWS directive directs CM-RTS to distribute the components of an array in the specified dimension evenly across the processing elements or vector units. Using this feature, three main layout directives for different classes of data structure used in the iterative PCG solution procedure are summarized below.

The unassembled element stiffness matrices k_e are stored in array ELEM_STIFF(DOF_ELEM, DOF_ELEM, NUM_ELEM) where DOF_ELEM is the degree of freedom per element (6 for space truss elements and 12 for space frame elements) and NUM_ELEM is the number of elements in the structural model. The layout directive for this type of data structure is ELEM_STIFF(:SERIAL,:SERIAL,:NEWS). With the first two dimensions made serial, the entire element stiffness matrix of an element is mapped to the same vector unit. Since the last dimension is NEWS, the element stiffness matrices are distributed evenly across the four vector units of the node by CM-RTS.

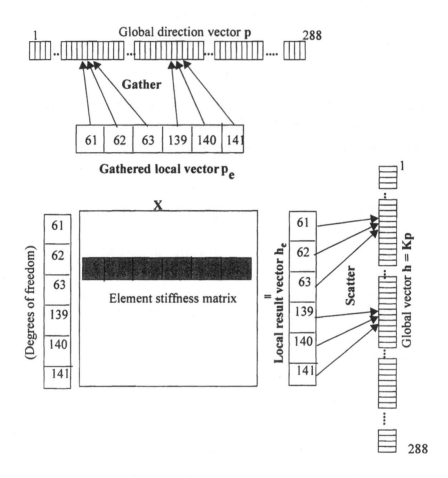

**Figure 6-5. Gather-and-scatter operations in
concurrent matrix-vector product**

The element-level unassembled vectors such as the element
connectivity array are stored in an array of the form CONNEC-
TIVITY(DOF_ELEM, NUM_ELEM). The connectivity array is
used as a pointer to the global assembled nodal vectors in gather-
and-scatter operations. The compiler directive (:SERIAL,
:NEWS) ensures even distribution of the array across four vector

units. The local element-level components of the global direction vector (p_e) and temporary results of the matrix-vector multiplication (h_e) are also laid out in a similar fashion.

The assembled nodal vectors (such as nodal displacement variable d) are stored in a one-dimensional vector such as U(DOF_TOTAL) where DOF_TOTAL is the total degree of freedom of the structure. The CM-RTS automatically distributes the assembled vectors across the processing elements.

With the above data layout, the element connectivity array, element stiffness matrix, temporary result of element-by-element matrix-vector product (h_e), and the local component of p (p_e) of the global assembled direction vector (p) all reside in the local memory of the same vector unit. Thus, once the local components (p_e) have been gathered from the global assembled vector, the matrix vector multiplication for all the elements can proceed simultaneously without any need for interprocess communication. Appropriate CM Fortran constructs or an assembly language routine from CMSSL can be used to perform this step concurrently.

6.5.1 Scatter-and-gather operations

The patterns of communication in scatter-and-gather operations are identical in each iteration of the PCG algorithm. Thus, the path of all the data elements being moved in a scatter or gather operation remains the same as long as the element connectivity does not change. The CMSSL routines *sparse_util_gather_setup* can be used to save the trace of communication activities involved in a typical gather operation. In the subsequent PCG iterations the same trace is used repeatedly for efficient gather communications. A similar routine exists for the scatter operation. It should also be noted that the data movement is within the nodes but across the vector units. Communication between the vector units of the same node which involves a memory-to-memory data transfer within the node is much faster than the node-to-node communication as required in the global execution model.

6.6 Examples and Results

The concurrent algorithm presented in this chapter has been applied to optimization of three highrise steel building structures. In all the examples a mutation probability of 0.0005 is used. Two-point crossover is applied and the number of bits used for encoding of the cross-sectional areas is 16. Other parameters such as population size (q), number of design iterations, and bounds for cross-sectional areas vary with each example and are given in Table 6-1. The general data for the three examples such as the number of elements, number of nodes, stress constraints, and the number of design groups are also given in Table 6-1. To improve the parent selection in the crossover operation of the genetic search, an improved stochastic remainder selection procedure has been used (Goldberg, 1989) instead of the roulette wheel selection method used in Chapter 5.

6.6.1 Example 1: 50-story space structure

The 848-element space truss structure (Figure 5-7) models the exterior envelope structure of a 50-story steel highrise building structure. More details about this structure along with the procedure used to compute horizontal loads can be found in the previous chapter (Section 5.4.3). In this example, the actual provisions of the AISC ASD code (AISC, 1989) for design of members in tension and compression have been utilized. For compression members, the buckling constraints are taken into account. The minimum weight obtained after 40 design iterations is 110.20 MN (compared with a minimum weight of 107.10 MN obtained using the distributed GA presented in Chapter 4). Due to the improved parent selection procedure, convergence is reached in a smaller number of design iterations (40 in this chapter compared with 60 reported in Chapter 5.)

6.6.2 Example 2: 35-story tower space structure

This example models the outer envelope of a 35-story space tower structure (Figure 5-9). More details about this structure (such as information about loading and constraints) can be found in the previous chapter (Section 5.4.4). The minimum weight obtained using the concurrent GA presented in this chapter after 40 iterations is 1.80 MN, which is slightly lower than that reported in the previous chapter using the distributed GA. It should be noted that design optimization results using a GA are not exactly reproducible due to the random nature of the search and machine-dependency of the random number generators.

Table 6-1 Summary of data for example structures

	Example 1	Example 2	Example 3
	50-story space truss structure	35-story space truss structure	147-story space truss structure
No. of elements	848	1262	4016
No. of nodes	224	324	817
Allowable tensile stress σ^u (MPa)	204.0 $(0.60F_y)$	170.0	204.0 $(0.60F_y)$
Allowable compressive stress, σ^l (MPa)	As per AISC ASD Specifications (Section E2)	−170.0	As per AISC ASD Specifications (Section E2)
Unit weight of material (kN/ m^3)	77.04	27.14	77.04
Young's modulus of elasticity, E (GPa)	199.95	69	199.95

Table 6-1 continued

	Example 1	Example 2	Example 3
	50-story space truss structure	35-story space truss structure	147-story space truss structure
Range of cross-sectional areas (cm^2)	0.645–2580	0.645–258	0.645–2580
Number of design groups (M)	67	72	251
Size of population (q)	256	256	512
No. of design iterations (s)	40	40	50
Minimum weight obtained using concurrent GA	110.20 MN	1.80 MN	882.7 MN
Minimum weight reported by other approaches	107.10 MN (Chapter 5)	1.87 MN Adeli and Cheng (1994b)	new example

6.6.3 Example 3: 147-story tower space structure

This example models the outer envelope of a 147-story monumental structure with a height of 564.06 m shown in Figure 6-6. The structure consists of 4016 truss members and 817 nodes. The 147-stories are divided into 49 modules. The cross-bracings in the lower 48 modules cover three stories. The height of the lower 48 modules is 11.43m. Two hundred and fifty-one design parameters are used to represent as many groups of members in this structure. The displacement constraints are given as 2.25m (±88.83 in.) at the top of the structure in all three directions (equal to 0.004 times the height of the structure). The loading on the structure consists of horizontal loads acting on the exterior nodes of the space structure at every three floors. The horizontal loads at each node are obtained from the Uniform Building Code's (UBC, 1994) wind loading. The value of C_q for the inward face is 0.8 and for the leeward face is 0.5. It is assumed the structure is located in a region with a basic wind speed of 112.63 Km/h (70 mph) resulting in a wind stagnation pressure of 603.53 N/m^2 (12.6 psf). The importance factor I is assumed to be one.

For this largest example structure used in this chapter, a large population size of 512 strings has been used. The design iteration history for this example is shown in Figure 6-7. A minimum weight of 882.7 MN is obtained for this example after 50 iterations.

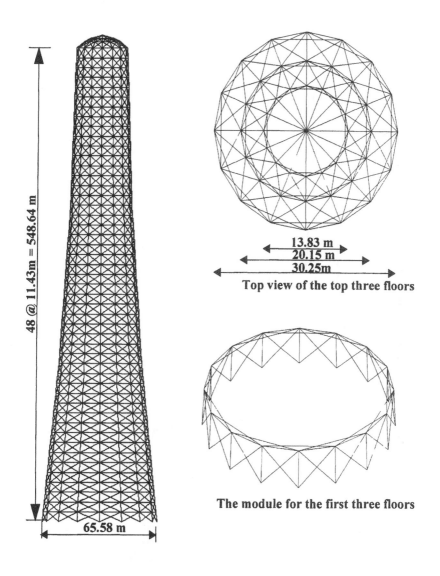

48 @ 11.43m = 548.64 m

65.58 m

13.83 m
20.15 m
30.25m

Top view of the top three floors

The module for the first three floors

Figure 6-6. Example 3 (147-story space tower)

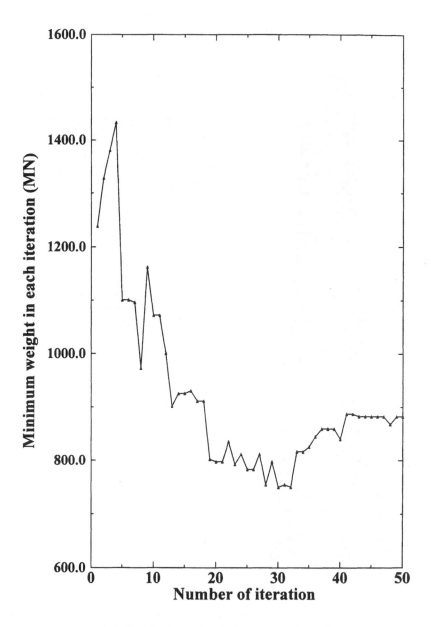

Figure 6-7. Variation of minimum weight for the 147-story space tower (Example 3)

6.7 Performance Results

Performance data on CM-5 can be obtained using the parallel debugging and performance analysis tool, Prism, or the CMMD timer library routines (Thinking Machines, 1993a&b). The latter is more accurate for time estimates of the critical code segments and has been used in this chapter. For the host/node model of computing, however, the CMMD timer library measures the busy time (elapsed time minus idle time) for the nodes only. It is not possible to measure the busy or the elapsed time for the host accurately. The ordinary CMOST library routine *gettimeofday* has been used to measure the wall-clock time. Since there is only one host and many nodes in partition and most of the computational work is performed on the nodes, the timing facility used to measure the performance is quite sufficient. Each node collects the timing data independently. At the end, all the results are combined at the host through a global reduction operation. CM Fortran compiler version 2.1.1-2 and CMOST version 7.3-3.3 have been used in this chapter.

In the following discussions, elapsed time refers to the overall processing time required to execute the program on a given partition of CM-5. This time includes busy time (when the host or a node is executing instructions) and idle time. It should be noted that since the CM-5 partition is used in a time-shared environment, the elapsed time is normally less than the wall-clock time.

6.7.1 Overall processing time

The overall processing times (elapsed times) for optimization of three example structures on a 32-node partition are presented in Table 6-2. Optimization of the structure of Example 1 takes about 1.13 hours on a basic CM-5 32-node partition. On a 256-

node partition this elapsed time is reduced to 16 minutes. For the largest structure, Example 3, the elapsed time on the 32-node partition is about 18 hours. The same structure can be solved on a 512-node partition in 1.15 hours. Thus, with a suitable choice of partition size the optimum solution for a large structure can be found in a reasonable amount of time.

Table 6-2 shows that the fitness function evaluation is by far the most time-consuming part of the GA-based structural optimization algorithm. On a 32-node partition it constitutes about 80% of the total elapsed time for the smallest Example 1 and increases to about 90% for Examples 2 and 3.

Table 6-3 presents a comparison of elapsed time for Example 2 on the Connection Machine CM-5 with the elapsed time on the Cray Y-MP 8/864. The elapsed time on a CM-5 64-node partition is comparable with that of Cray Y-MP 8/864. The processing time required by the concurrent algorithm presented in this chapter on a CM-5 256-node partition is approximately one-third of that reported by Adeli and Cheng (1994b) on a Cray Y-MP 8/864.

6.7.2 Scalability

The small difference between the overall elapsed and busy times in Table 6-2 is due to the sequential population replication on the host. The processing time required for the population replication remains relatively constant as the number of processors is increased. The computationally intensive fitness function evaluation is highly scalable. Since the population replication constitutes only a small fraction of the overall elapsed time, the result is a highly scalable algorithm. This important conclusion is clearly demonstrated in Figure 6-8. This figure shows the speedup obtained for the three example structures relative to the basic 32-node partition as we increase the number of processors to 256.

Table 6-2 Processing times on 32-node CM-5

Steps	Example 1	Example 2	Example 3
Overall elapsed time (hr)	1.13	2.34	17.99
Overall busy time (hr)	1.08	2.29	16.41
Fitness function evaluation (hr)	0.91	2.13	15.36
Population replication (s)	155.04	531.67	3616.72
Communication (s)	4.69	1.82	8.28
Disk I/O (s)	11.63	21.24	136.9

Table 6-3 Comparison of results for Example 2 on a CM-5 and Cray Y-MP 8/864

Characteristics	Adeli and Cheng (1994b)	Present Study
Population size	140	256
Number of design iterations	50	40
Minimum weight	1.87 MN	1.80 MN
Elapsed times	1.27 hr Cray Y-MP 8/864	32-node: 2.34 hr 64-node: 1.23 hr 256-node:0.376 hr

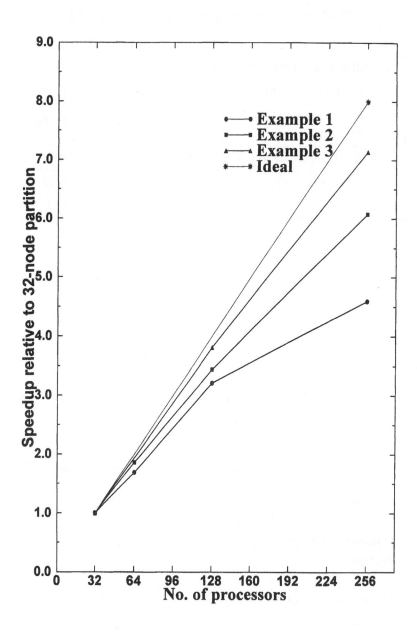

Figure 6-8. Speedup for example structures relative to 32-node partition

To improve the scalability of the concurrent algorithm even further, the population replication step should also be parallelized. In this case, the suitable execution model would be the global/local model. In each design optimization iteration, after the fitness function is evaluated using the local (node-level) execution model, the population is replicated using the global model of execution. A suitable data mapping strategy for the data parallel population replication can be developed. Bits in the strings (chromosomes) of a population can be distributed across all the vector units in the partition and genetic operations (such as crossover and mutation) can be performed on them concurrently. However, as stated earlier in this chapter, the global-local model of execution is not currently supported by the Connection Machine.

6.7.3 Iterative PCG solver

The iterative PCG solver forms the computational kernel of the fitness function evaluation. The distribution of average busy time required in various steps of the solution of the system of linear equations, Eq. (6-1), is given in Table 6-4. Most of the floating point operations are performed in the matrix-vector multiplication step ($72 * N_E$ FLOP for space structures consisting of truss elements). This step is preceded by a gather operation and followed by a scatter operation, both of which involve local communication between the vector units of the same node. The number of 64-bit floating point (FP) words communicated in this step is 6 times the number of elements in the space truss structure. Computations involved in other operations of a PCG iteration (such as dot products) are relatively small (approximately 6 times the number of degrees of freedom of the structure). However, they have been included in the performance evaluation of the iterative PCG solver presented Table 6-5.

Local communication bandwidths (the number of FP words communicated times 8 divided by the busy time) between the vector units during the scatter and gather operations are also given in Table 6-5. The observed bandwidth is in the range of 10–12 Mbytes/s for both scatter-and-gather operations. The theoretical peak bandwidth between vector units on the same chip of a node is 20 Mbytes/s and the peak bandwidth for communication between vector units on different chips of a node is 10 Mbytes/s. Thus, the observed bandwidth is quite impressive considering the irregular nature of communication in scatter-and-gather operations.

The FLOP rate for the PCG algorithm is given in Table 6-5. It is noted that the MFLOP rate increases with the size of the structure. This rate is almost independent of the processing nodes in the partition. For the largest, Example 3, at a rate of 4.71 MFLOPS per processing node, a combined rate of 2.4 GFLOPS is observed on a 512-node CM-5.

**Table 6-4 Processing time distribution (in seconds) for the
iterative PCG solver on one node of CM-5**

Steps	Example 1	Example 2	Example 3
Total busy time required for one solution	3.71	23.515	71.825
First iteration and setup	0.85	1.215	4.31
PCG iterations	2.85	22.26	67.48
Scatter operation	0.63	5.24	17.09
Gather operation	0.71	5.25	20.31
Matrix-vector multiplication	1.367	10.67	28.00
Other operations	0.108	0.825	1.515

**Table 6-5 Performance rate and communication
bandwidths for iterative PCG solver**

Steps	Example 1	Example 2	Example 3
Average number of PCG iterations	175	993	1046
FLOP per PCG iteration (72 * number of elements + 6 * degrees of freedom)	65088	96692	303858
FP words scattered/ gathered in one PCG iteration (6 * number of elements)	5088	7572	24096
Average CM busy time per PCG iteration (milli-seconds)	16.28	22.41	64.51
MFLOPS per node	3.99	4.31	4.71
Average CM-busy time per scatter operation (milli-seconds)	3.59	5.26	16.33
Average CM-busy time per gather operation (milli-seconds)	4.06	5.29	19.41
Scatter bandwidth (Mbytes/s)	11.32	11.51	11.80
Gather bandwidth (Mbytes/s)	10.00	11.43	9.93

7 Concurrent Animation of Seismic Response of Large Structures in a Heterogeneous Computing Environment

7.1 Introduction

The finite element method can be used for computation of the dynamic response of large complicated structures. However, computation of time-history seismic response is a very time-consuming process and animation of response in real time requires high-performance computing resources. For example, to animate the response of a structure in real time for an earthquake of 30 seconds duration, the entire simulation should take no more than 30 seconds. As such, an effective strategy needs to be devised for the animation of response of large structures. Consider the following facts:

- Most large mainframes and supercomputers do not have advanced graphic capability. This means that while dynamic analysis will be quite fast on them, animation will be too slow.

- Due to the numerous complexities involved in graphic pre-
  and post-processing, object-oriented programming (OOP) is a
  desirable choice for the programming model. However, most
  supercomputing environments, due to their emphasis on high-
  performance computing using FORTRAN, do not support an
  advanced interactive development environment for C++ or
  other object-oriented (OO) languages.

- Using an expensive supercomputer for graphic processing
  may not be the best use of its computing power.

- Most supercomputing sites have restrictions on the maximum
  duration for which an interactive task can be run.

- Existing finite element dynamic analysis programs (usually
  written in FORTRAN) do not have animation capability. They
  normally have a stand-alone graphics interface written in C or
  C++.

Thus, in this chapter we present an effective strategy for real-
time graphic animation of earthquake response of large structures
that can be incorporated into a general-purpose finite element
dynamic analysis system in a distributed or parallel/vector computing environment. We present a two-pronged strategy to
develop an object-oriented graphic interface on a workstation
combined with computations of seismic response on a supercomputer or a mainframe computer. Communication and synchronization constructs from the software library PVM (Parallel Virtual
Machine) (Geist et al., 1993) are used for coordination between
the heterogeneous applications running on heterogeneous Unix
machines (CRAY YMP 8/864 supercomputer and IBM RISC/
6000 workstation).

The approach, wherein two different applications and
machines work on separate parts of problems making the best use
of their specific capabilities, provides a cost-effective strategy to
combine the computing speed of a supercomputer with interactiv-

ity and graphical capability of a workstation. While the CPU-intensive seismic response calculations are performed on a super-computer, the animation of the deflected profile of the structure proceeds simultaneously on a relatively inexpensive workstation. Developed in an X Window environment, the graphic animation system is based on a hidden line removal scheme combined with an object-oriented knowledge-base system for graphical modeling of the structure.

An effective communication strategy for overlapping communications with computations has been developed to minimize the effect of network delays. Seismic analysis is based on the mode superposition method used in conjunction with the subspace iteration method for extraction of mode shapes and the frontal solver for the out-of-core solution of linear system of equations. A vectorized 3-point integration procedure is used for transient response computation by the Duhamel integral method.

7.2 Dynamic Analysis of Structures

The dynamic equation of motion for a structure subjected to dynamic loading is (Bathe and Wilson, 1982):

$$M\ddot{U} + C\dot{U} + KU = R(t) \qquad (7\text{-}1)$$

where U is the time-dependent vector of displacements, $\dot{U}$ is the vector of nodal point velocities (first time derivative of U), $\ddot{U}$ is the nodal acceleration (second time derivative of U), R is the vector of time-dependent loading, M is the assembled mass matrix of the structure defined as (e refers to summation over elements):

$$M = \sum_e \int_{V^e} \rho^e N^{eT} N^e dV^e \qquad (7\text{-}2)$$

K is the global stiffness matrix of the structure defined as:

$$K = \sum_e \int_{V^e} B^{eT} D^e B^e dV^e \qquad (7\text{-}3)$$

and **C** is the damping matrix of the structure.

Mathematically, Eq. (7-1) represents a system of linear differential equations of the second order and is usually solved either by direct integration or the mode superposition method, taking advantage of symmetry and diagonal dominance of **K**, **C**, and **M** matrices. In the direct integration approach, Eq. (7-1) is numerically integrated using a step-by-step procedure. The total number of operations required in complete integration is approximately $\alpha N m_K s$ where N is the number of degrees of freedom of the structure (order of the matrix **K**), and m_k is the semi-bandwidth of **K** (Bathe and Wilson, 1982). Coefficient $\alpha \geq 2$ depends on the characteristics of matrices and s is the number of time-steps. Thus, in the direct integration method, the cost of analysis is directly proportional to the number of time-steps (the time-step used in seismic response analysis is usually 0.01 seconds). Hence, direct integration schemes are effective for response calculations for a short duration only (a few time-steps). Further, for large structures, computation involved at each time-step is substantial. This computation needs to be performed within 0.01 seconds for animation to be perceived as a continuous real-time sequence which is not feasible even on large supercomputers today. As a result, mode superposition is the preferred method of linear dynamic analysis for real-time animation of a large structure.

In the mode superposition method following linear transformation is applied to displacement U:

$$U(t) = \Phi Y(t) \tag{7-4}$$

where Φ is the square matrix of mode shapes, and $Y(t)$ is the time-dependent vector of normal coordinates.

The major computational task in mode superposition analysis is the computation of eigenvalues and eigenvectors. For linear dynamic analysis of structures, they need to be computed only once. Further, only the first few modes of vibrations contribute to the response of the structure and hence need to be included in the analysis. Simulation of different earthquake records (accelerograms) can be performed using the same eigenvalue and eigenvector solution.

7.3 Subspace Iteration Method for Eigenvalue Extraction

The subspace iteration method is the most efficient method for an out-of-core solution of the eigenvalue problem that arises in dynamic analysis of large structures. Since the K and M matrices for large structures do not fit into the main memory of even the largest computers, out-of-core solvers become a necessity. The subspace iteration method has been developed to minimize the operation on out-of-core disk storage. It can he summarized by following two steps:

Step 1: Establish p starting iteration vectors, $p > n$, where n is the number of eigenvalues and vectors to be calculated.

Step 2: Use subspace inverse iteration and Ritz analysis on the p
vectors to extract n eigenvalues and eigenvectors.

Details of subspace iteration method for mode extraction can
be found in Bathe and Wilson (1982).

7.4 Graphic Animation in a Heterogeneous Environment

The major steps in the mode superposition analysis are:

1) solution of eigenvalue problem,

2) solution of uncoupled equations of motion, and

3) the superposition of modal responses using Eq. (7-4).

In our computational model the first two steps are performed
on a remote supercomputer (henceforth referred to as the server
machine) and the final step is performed on a Unix workstation
(referred to as client machine). This has been done in accordance
with relative performance rating of the two machines and the cost
of communication involved.

In dynamic earthquake analysis only the first few modes of
vibration usually need to be considered. The deflected profile of
the structure in step 3 can be computed from a small number of
modal coordinates $Y(t)$ if the mode-shape matrix is made available to the client machine. This involves communication between
the server and the client machine. However, this communication
overhead is more than offset by the gain in performance by overlapping communications with computations. A summary of the
steps performed by both the client and server processes and their
coordination is given in Table 7-1. The tasks performed by each
process are shown in their respective column.

Table 7-1 Coordination between heterogeneous tasks for dynamic analysis of structures

Step	Client process on Unix Workstation (IBM RS/6000)	Server process on Cray YMP 8/864 supercomputer
1	Read mesh data. Create the knowledge base of connectivity (as described in Section 3.2). Save the knowledge base on out-of-core storage.	Obtain mesh data. Perform subspace iteration and save natural frequencies and mode shapes on the hard disk.
2	Read the knowledge base (created in Step 1) from the hard-disk. Get the user preferences for viewing angle, distance, and magnification factor f for displaying nodal displacements. Spawn the server process on the remote supercomputer using *pvm_spawn* library call.	Server process started. Read earthquake data (accelerogram) from the hard disk.

Table 7-1 continued

Step	Client process on Unix Workstation (IBM RS/6000)	Server process on Cray YMP 8/864 supercomputer
3	Enter an X Windows event loop. If an "expose" event is received (this event is received when the user reconfigures the windows dimensions), plot a new hidden line filtered image of the model (Section 3.3.1). This is the static image that remains on screen throughout the simulation.	Enter an infinite loop and wait for a message from the client process (until client process sends a termination request).
4	Start a loop over the time-steps. Send a request to the server process to start animation.	Request received from the client process to start animation. Start the loop over time-steps.
5	Receive modal response Y and compute the deflections U for visible nodes only using Eq. (7-4).	Evaluate Duhamel's integral and send computed modal values, Y, to the client process.

Table 7-1 continued

Step	Client process on Unix Workstation (IBM RS/6000)	Server process on Cray YMP 8/864 supercomputer
6	Compute the deflected node position coordinate (X', Y', Z') as: $X' = X + fU_x$ $Y' = Y + fU_y$ $Z' = Z + fU_z$ where (X, Y, Z) is the original node position coordinate, U_x, U_y, U_z are the components of deflection in each coordinate direction and f is the user-specified magnification factor. Transform deflected node positions. Erase old deflected profile. Plot new deflected profile. Go back to the top of the time-step loop (Step 5) until maximum number of time-steps has been reached.	Go back to the top of the time-step loop (Step 5) until maximum number of time-steps has been reached.
7	If user wants to quit, send 'STOP' signal to the server process. Else, go back to the beginning of the X Window event loop (Step 3).	If 'STOP' signal is received from the client process, then end the execution. Else go back to the beginning of the event loop (Step 3).

As Table 7-1 shows, the client process on the Unix Workstation operates as the front-end to the client-server graphic animation system. A flow-diagram of this process is given in Figure 7-1. First, the client process accepts the user's selection for viewing angle, distance, focal point and magnification factor (f) to generate a static view of the structural model. It then iterates in a loop over time-steps, accepting the modal deflection from the server process at each time-step. It computes the deflected profile of the structural model and superimposes the dynamically generated view over the static view in the display window. The static view is generated by performing the hidden line removal as outlined in Section 3.3.

At the core of our graphic animation system is the class of object-oriented data structures described in Chapter 3. Object-oriented programming methodology makes it easy to extend the existing set of classes (such as *Node* and *Element*) developed in Chapter 3 with new methods for computing and displaying the profile of the structural model dynamically. For example, we added a method to the *Node* class to compute the deflected location from the old location given the deflection and magnification factor. The new method reuses the existing implementation of the *Node* class to compute the new information needed for dynamic analysis.

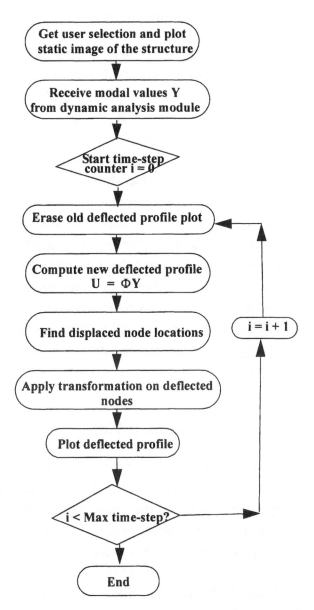

**Figure 7-1. Flow diagram for the front-end of the
graphic simulation module**

7.4.1 Event-driven X Window interface

To display the result of graphic animation we use the X Window system, an industry standard, vendor-independent, network transparent graphics environment for Unix systems. An application developed using the X Window system is event-driven. For example, if the user relocates or resizes the window, X window manager sends a set of events to the application notifying it of the user's action. In the graphic animation system we use this event mechanism to control the interaction with the user. For example, if the user hides the display window behind a stack of other windows and then exposes it, the application performs the following steps in response to the X Window event:

Step 1: Clear the window and draw the static image.

Step 2: Send a message to the remote server process to start computing new set of modal displacements.

Step 3: Receive the modal response data and plot the animated mesh view.

7.4.2 Parallel Virtual Machine (PVM)

In our implementation of the dynamic animation system the software library PVM (Geist et al., 1993) has been used to provide communication between the client (running on a IBM RS/6000 workstation) and the server process (on the CRAY YMP 8/864 supercomputer). It may be noted that the CRAY YMP 8/864 supports 8-word floating point numbers whereas workstations are 32-bit machines. PVM automatically converts data between different floating point representation schemes and is ideal for communication between machines of different architectures and processor types.

7.4.3 Effective communication strategy

In this section we present effective communication strategies between the two heterogeneous machines: a Unix Workstation IBM RS/6000 and a vector supercomputer CRAY YMP 8/864. The CRAY YMP 8/864 performs the dynamic analysis and Workstation performs the graphical animation. Our ultimate aim is to plot the deflected profile of the mesh at every time-step on the workstation screen. Since the ground acceleration data is usually available at a 0.01-second time interval, it is desirable that the entire computation and plotting of image take less than this time. This appears to be a daunting task at first. We exploit the following factors and strategies:

First, the separation of the transient part from the static part. That is, only the computations relating to time-varying quantities are performed at each time-step. All other transformations which do not change over time are performed once only.

Second, time lost in communication delays is overlapped with useful computation at both client and server machines. This is due to the asynchronous nature of the communication involved between the client and the server process.

Third, since cost of communication is much more than the cost of computation, minimum communication is used. For example, the remote server process that performs the dynamic analysis can either send the deflections for the N total degrees of freedom (DOF) or it can send just the modal response and let the client process compute the actual deflections U from Eq. (7-4). Since the size of Y vector (equal to the number of modes) $n \ll N$, it is advantageous to send only the values for Y if the mode shape matrix Φ is transferred from the server process to the client process. To evaluate the effectiveness of this strategy, let us assume $n = \alpha N$, where α is a coefficient (to be determined later) that will determine the effectiveness of our strategy. In that case the total number of floating point (FP) words communicated in one time-step is

$$C_1 = (\alpha N)s + q\alpha N \qquad (7-5)$$

If, instead, the deflections are computed by the server process and then communicated to the client process, the total number of floating point words communicated in each time-step will be $C_2 = Ns$. For $C_1 < C_2$, we need

$$\alpha Ns + \alpha N^2 < Ns \qquad (7-6)$$

which means

$$\alpha < \frac{s}{(s+N)} \qquad (7-7)$$

For example, if we consider integration over 2400 time-steps of a 3000-DOF finite element model, we need $\alpha < 45\%$. Since in most simulations we will be using only a small percentage of total mode shapes, our strategy saves the cost of communication by at least a factor of about two in most cases.

7.5 Example

As an example, the 100-story building structure described in Section 3.4.2 and shown in Figure 3-11 is subjected to the 1940 El Centro earthquake record shown in Figure 7-2. A screen-shot of the animated profile generated by the animation system is shown in Figure 7-3. This figure demonstrates that our approach to graphic animation of dynamic response is effective in handling very large structures.

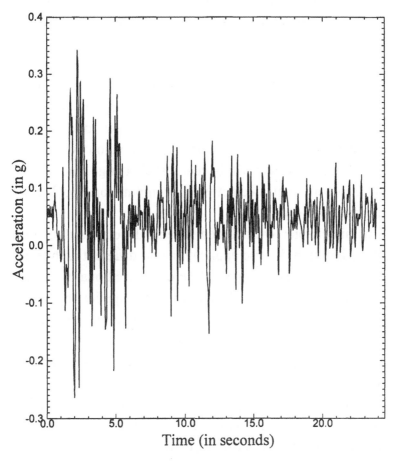

Figure 7-2. Variation of ground acceleration with time for the 1940 El Centro earthquake

7.6 Final Comments

Engineers currently use a large array of application programs for their diverse applications such as analysis, design, and visualization. Conversion of data between different systems is often not a smooth process. One application needs to be terminated and another has to be started, often on a different machine. Sometimes the volume of data transfer can be too large and in such cases no amount of data compression and universal file specification can help. Animation is one such case. A complete response history of a large model (10,000 DOF) subject to an earthquake acceleration of only 30 seconds duration (3000 time-steps at 0.01-second intervals), consists of approximately 30 Megawords of data transfer between dynamic analysis and the animation system. Obviously, animation effect will be lost if such a large amount of data need to be communicated between a computational back-end and a graphic front-end.

In this chapter we outlined a general purpose data-transfer strategy between diverse applications on different architectures, based on message-passing in place of conventional file transfer. Our approach maintains interactivity of the applications and provides for smooth communication and cooperation between different tasks. A common user interface based on the X Windows system drives applications running on architectures as different as a PC and a CRAY supercomputer.

With the arrival of high-speed networks and vendor-independent communication library like PVM, this approach of heterogeneous computing in a network environment will become a new standard. We may see emergence of standard message-passing interface and its use by developers of analysis and design software in place of file format specifications.

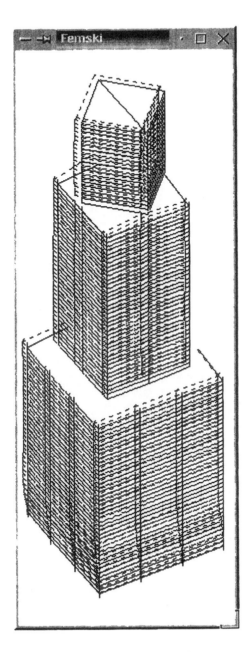

**Figure 7-3. 100-story highrise building structure
subjected to El Centro earthquake**

8 New Directions and Final Thoughts

8.1 Introduction

In this chapter we present new directions and our views on contentious issues that engineers face today in the design and development of distributed solutions for large-scale problems in analysis, design and visualization. We try to put the relevance of the algorithms and methods presented in this book in the context of emerging trends in technologies. Selection of CAE solutions for engineering problems often involves carefully weighing the different solutions available and then making an intelligent decision about the most suitable choice for the given problem. The size and nature of the problem play an important role as does the availability of a hardware platform to implement that solution. For example, most small enterprises cannot afford a large super-computer. Thus, the choice of a network of computers (Unix Workstations or Windows NT PCs) becomes a pragmatic one. Similar decisions need to be made about different choices available for the solution method for systems of linear and non-linear equations (direct versus iterative methods), design optimization methods, operating systems (Unix vs. Windows NT), visualization methods (such as VRML), and communication library (RPC, CORBA, PVM or MPI).

201

8.2 Direct Versus Iterative Solvers

In the past decade or so the debate has been raging among researchers regarding the choice of a solution method for solving the system of linear equations arising in static finite element analysis:

$$\mathbf{Ku} = \mathbf{f} \qquad\qquad (8\text{-}1)$$

The discretization level and, hence, the accuracy of a FE model are often determined by the memory requirement of the system of linear equations (8-1). Even though memory is becoming increasingly inexpensive, the sophistication of finite element analysis procedure has grown with time, too. Engineers' demand for greater accuracy in the analysis process has led to finer discretization and the introduction of memory intensive procedures such as the p-version finite element methods (Rank and Babuska, 1987).

Most of the commercial, sequential finite element programs use some variant of the frontal solver, a direct solution method based on the Gaussian elimination scheme (Irons, 1970). Based on a careful *a priori* analysis of structural connectivity in the pre-front phase, the frontal solver does an intensive 'housekeeping' to keep the memory requirement small. However, the algorithm is sequential in nature and high I/O frequency further restricts parallelizibility (Lesoinne and Farhat, 1991) even though multifrontal solution procedures have been proposed (Duff and Reid, 1983).

Most of the parallel algorithms proposed in the literature for solution of Eq. (8-1) are based on the direct solution method using a banded or skyline storage scheme (Duff et al., 1986; Ashcraft et al., 1987; Farhat and Wilson, 1988). The skyline storage scheme is more economical when it comes to memory requirement. However, it suffers from a serious drawback due to problems associated with fill-ins: triangular factors of **K** have many more nonzeros than the original matrix itself. For large-scale three-dimensional structures, the fill-in required grows proportionally to $N_D^{1.5}$ and the solution time degrades proportionally to N_D^2 or worse (N_D is the total number of degrees of freedom for the structure).

Considerable expertise has been invested in designing sparse storage schemes and out-of-core assembly and elimination processes to keep the memory requirement small and solution time manageable (Sarma and Adeli, 1995, 1996). However, the use of slow out-of-core (secondary) storage becomes unavoidable for large problems. This limits their parallelizibility and suitability in an environment of a cluster of workstations. It is in light of these limitations of direct solvers that we employed an iterative solver in our distributed algorithms. Iterative solvers have a small memory requirement since the global stiffness matrix need not be stored and computation can be done on an element-by-element basis. Element-by-element computation also lends itself to a high degree of parallelization (Johan et al., 1992, 1994; Johnsson and Mathur, 1990). However, as the subsequent section illustrates, advances in preconditioning methods have blurred the lines between direct and iterative solvers. Modified versions of direct solvers are used more and more in the preconditioning phase of an iterative solver today.

8.3 Hybrid Preconditioner for Large-Scale Problems

In our distributed PCG algorithms presented in Chapters 2 and 3 we used a diagonal preconditioner for its simplicity and ease of implementation. This is by no means a limitation of our algorithms. In fact, any other preconditioner can be employed. For structural models consisting of distorted elements or elements with anisotropies, convergence of an iterative method can be improved substantially with the design of an appropriate precon-ditioner. *A priori* estimates of convergence of iterative solvers based on the condition number of the preconditioned system of equations exist and an adaptive approach can be used to develop an appropriate preconditioning scheme for progressively more difficult problems. The preconditioned system of equation:

$$\mathbf{H}^{-1}(\mathbf{K}\mathbf{u} - \mathbf{f}) = 0 \tag{8-2}$$

can also be thought of as a solution for the auxiliary system of equation:

$$\mathbf{H}\mathbf{u} = \mathbf{r} \tag{8-3}$$

where $\mathbf{H}$ is the preconditioning matrix. Between the extremes of the largest value of $\mathbf{H} = \mathbf{K}$ (effectively meaning a direct solution method) and $\mathbf{H} = \mathbf{I}$ (meaning a pure CG method with no preconditioning), there are trade-offs in the choice of the preconditioning matrix $\mathbf{H}$. In other words, an approximation to $\mathbf{K}$ is used for $\mathbf{H}$.

Preconditioning matrix $\mathbf{H}$ can be generated by using an adap-tive approach based on element-level transformations (Mandel, 1993) or a global search using heuristics based on relative magni-tudes of entries in the global stiffness matrix (Suarjana, 1998; Ajiz and Jennings, 1984). Apart from a smaller memory require-ment than the original $\mathbf{K}$ matrix, preconditioning matrix $\mathbf{H}$ gener-ated by these methods has many special properties which make it easier to factorize using the Cholesky factorization ($\mathbf{H} = \mathbf{L}\mathbf{L}^{T}$) scheme. The domain decomposition method presented in

Chapter 2 can be used to factorize the preconditioning matrix in parallel on a network of workstations and the PCG method can be applied subsequently to the Cholesky factor of **H** matrix on the subdomain level. For large problems with over 1 million degrees of freedom, a hybrid iterative solver outperforms the direct skyline solver by a factor of more 40 in CPU time and more than 16 in disk space usage and is shown to handle elements with very bad aspect ratios of up to 1:500 (Mandel, 1993).

Our general distributed algorithms can accommodate a choice of solution methods (direct or iterative) and preconditioners. The selection of the most preferred solution method or preconditioning scheme should be based on the size and complexity of the given structural problem and the available computer hardware. An approach that restricts the users to only a direct or iterative solver or only one type of preconditioning will be of limited practical interest.

8.4 Operating Systems War: Unix versus Windows NT

Recent emergence of the Windows NT operating system as a cost-effective alternative to Unix offers engineers an interesting choice of operating system. The technical workstation market has long been a Unix bastion. Unix workstations are known for their scalability and robustness and excellent development environment based on an "*open systems*" approach. However, Unix systems are based on a proprietary processor and hardware architecture and hence, cost substantially more than a typical personal computer running Microsoft's Windows operating system. With the advent of Intel's powerful Pentium Pro processor, inexpensive workstations running Windows NT operating system are gaining increasing popularity in the engineering marketplace.

Due to wide availability of 3rd party hardware components ranging from 3D graphics acceleration cards (for hardware acceleration of 3D API such as OpenGL and Direct3D) to high speed disk
storage systems, Windows NT workstations cost much less than
similarly equipped low-end Unix systems. To facilitate migration
of Unix applications, Windows NT supports many of the systems
and communications APIs (Applications Programming Interface)
such as RPC, CORBA and PVM that have long been available on
Unix. We have been able to port most of the algorithms presented
in this book to Windows NT relatively effortlessly using an
implementation of PVM available on Windows NT.

However, while there are efforts to enhance Windows NT's
capability using symmetric multiprocessing, scalability of Windows NT systems is currently limited. As a result Unix workstations are still indispensable for the largest structural engineering
problems. Further, several variants of Unix (in particular, the
Linux operating system) are now available on Intel platforms and
have been gaining popularity. "Unix on Intel" systems provide
the best of both worlds: open hardware architecture based on
Intel's processor and the open development environment of the
Unix operating system.

With the arrival of more powerful 64-bit processors in the
future, it appears that both "Unix on Intel" and Windows NT will
become more powerful and popular alternatives to traditional
RISC-based Unix Workstations. This means more powerful systems will be available to engineers at a lower cost in the future.
This will further enhance the price-to-performance advantage of
workstation clusters over traditional supercomputers since engineers will be able to create distributed computing environments
using inexpensive workstations and communications API such as
PVM. A side-by-side comparison of Unix and Windows NT is
presented in Table 8-1.

Table 8-1 : Comparison of Unix and Windows NT

Feature	Unix	Windows NT
GUI	Layered on top of OS Kernel. X Windows is the basic graphics subsystem, Motif is the GUI widget set and CDE is the look-and-feel standard based on X/Motif. X Windows is network-enabled, letting any GUI application be run remotely.	GUI is embedded in the OS kernel and there is no choice of GUIs. Windows 95 is the common look-and-feel. NT is based on a desktop paradigm meaning only one user can access a machine at a time. It does not support remote login.
System administration	Managed using low-level character-based scripts. Easy to access all administrative functions remotely. Supports user level quota for system resource usage.	Most administration tools are GUI-based. Cannot be managed remotely. Lacks scripting support. No user quotas available.
Scalability	Proven to scale from workstations to super-computers. 8 to 64-processor symmetric multiprocessing systems are common. Several choices for fail-safe clustering available.	Scalability limited but improving. While theoretically 32-processor symmetric multiprocessing is possible, most implementations are limited to 2 or 4 processors. Clustering standards are emerging.

Table 8-1 continued

Feature	Unix	Windows NT
Standard-ization	Based on POSIX standard and other open APIs. Many of the Unix APIs have become internet standards. However, various flavors of Unix have their own enhancements, thereby impairing portability of applications.	Controlled by a single vendor (Microsoft)-all versions of NT share the same APIs and system calls. Thus the software developer need not re-compile system for various flavors. This has facilitated development of 3rd party applications.
Cost	Each Unix vendor has its own processor and system architecture. Limited support of 3rd party hardware. Lack of competition in hardware components keeps systems cost high. Requires advanced skills to administer and use. Very few personal productivity applications available.	Single system API and driver model facilitates development of 3rd party hardware. Competitive 3rd party hardware add-on market lowers the system cost. Easy to administer and use meaning lower overall cost of administration. High productivity due to availability of personal productivity applications such as word processors.

8.5 VRML: 3D Visualization on the Internet

In Chapter 7 we presented a real-time visualization model based on the client-server approach for animation of dynamic response of structures subjected to earthquake forces. In our approach emphasis is on exploiting the computational power of a supercomputer for earthquake analysis coupled with animation on a graphics workstations. The client-server approach based on collaboration using synchronous communication helps avoid slow and inflexible file transfer protocols. However, in a loosely coupled asynchronous messaging environment (internet e-mail or file-transfer protocol, ftp), file-based collaboration becomes necessary.

There have been several efforts to develop open-file formats for exchange of engineering data for collaboration among engineers. While these formats have proven to be suitable for exchange of complex engineering data, lack of cooperation among vendors has thwarted their large scale acceptance. In recent years, Virtual Reality Markup Language (VRML) (Hartman and Wernecke, 1996) has emerged as a powerful medium for interactive exchange of information between online users on the world wide web (WWW).

A VRML model (also known as "world" in the internet community) blends 2D and 3D objects, animation and multimedia effect in a single, interactive medium. While most of the VRML applications are in the gaming and online shopping domain, we think VRML can be leveraged effectively for cooperation and exchange of graphic data among engineers, since VRML primitives are powerful enough to model complex engineering models.

Since VRML viewers are included with popular web browsers such as Microsoft Internet Explorer and Netscape Communicator, engineers can view and navigate a VRML *world* on most platforms, ranging from inexpensive PCs to powerful Unix workstations. The vendor-neutral, platform-independent nature of the VRML standard makes it an attractive alternative to conventional engineering file formats that require expensive viewer software. Figure 8-1 shows a turbine equipment being viewed in the Internet Explorer browser.

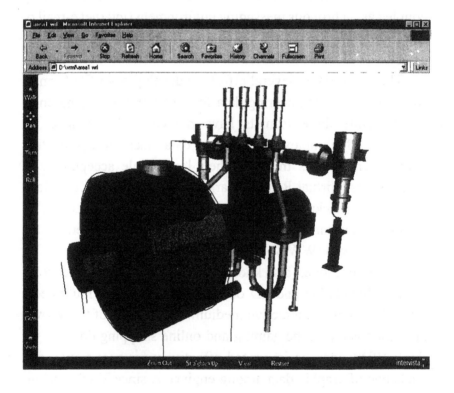

Figure 8-1. VRML model of a turbine equipment in Internet Explorer browser (Courtesy of Jacobus Technology, Inc.)

8.6 Network Infrastructure for Distributed Computing

With the rising popularity of the internet, there has been significant improvement in the networking infrastructure in most enterprises. A local area network based on the TCP/IP (Feit, 1996) protocol is now deployed in most organizations. This has given engineers a plethora of choices when it comes to selection of suitable components for a distributed computing infrastructure. The APIs available to create a distributed computing environment can be divided into three different categories. At the lowest level are inter-process communication protocols such as UDP (User Datagram Protocol), and TCP/IP. At the next higher level of abstraction are sockets which provide protocol-independent network interface services. Sockets hide the network-level details, and an application developed using sockets can run on many different types of networks using many different protocols. However, programming using sockets is still much like programming in assembly language. Users have to develop their own conventions for the exchange of names and arguments and provide their own marshalling of data.

At the highest level of abstraction are Object Request Broker's (ORBs) such as CORBA (Common Object Request Broker Architecture) from Object Management Group (Siegel, 1996; Orfali and Harkey, 1997; Vogel and Duddy, 1997), DCOM (Distributed Component Object Model) from Microsoft (Grimes, 1997) and RMI (Remote Method Invocation) from Sun (Farley and Farley, 1998). ORBs hide all details of distributed network programming. Users define an interface and implement that interface in their language of choice. The ORB vendors provide a stub compiler to automatically generate code that facilitates actual wire-level data transfer.

Questions may arise as to whether use of a low-level communication library such as PVM is prudent in light of all these developments happening in the internet community. PVM provides a higher level of abstraction than sockets without much loss of performance. PVM is a complete environment for development of distributed engineering applications at the right level of abstraction. Sockets are too low level for engineers and ORBs are too high level, involving a much higher communication overhead. PVM has been developed keeping the needs of engineers and scientists in mind (for example, it provides binding for FORTRAN). While ORBs simplify the task of distributed programming, applications developed using them are non-portable across different vendor ORBs, raising the cost of overall development. PVM is a public domain software available free of cost. It is stable and mature and widely available on most Unix workstations and supercomputers. With successful completion of current efforts to enhance visual application development environment for PVM, it appears that PVM will remain the method of choice for distributed application development in the engineering and scientific community.

8.7 Final Comments

With wide-scale acceptance of the internet in enterprise computing, distributed network infrastructure is rapidly growing in most organizations. This lowers the cost of distributed computing and provides a fertile ground for further development of methods presented in this book. Algorithms presented in this book can be enhanced easily to take advantage of emerging technologies such as hybrid preconditioners for iterative solution, lower cost Windows NT Workstation cluster, and VRML-based data visualization schemes.

References

Adeli, H., Ed. (1988), *Expert Systems in Construction and Structural Engineering*, Chapman and Hall, London.

Adeli, H., Ed. (1992a), *Parallel Processing in Computational Mechanics*, Marcel Dekker, Inc., New York.

Adeli, H., Ed. (1992b), *Supercomputing in Engineering Analysis*, Marcel Dekker, Inc., New York.

Adeli, H. and Cheng, N.T. (1993), "Integrated Genetic Algorithm for Optimization of Space Structures," *Journal of Aerospace Engineering*, ASCE, Vol. 6, No. 4, pp. 315-328.

Adeli, H., and Cheng, N.T. (1994a), "Augmented Lagrangian Genetic Algorithm for Structural Optimization," *Journal of Aerospace Engineering,* ASCE, Vol. 7, No. 1, pp. 104-118.

Adeli, H., and Cheng, N.T. (1994b), "Concurrent Genetic Algorithms for Optimization of Large Structures," *Journal of Aerospace Engineering*, ASCE, Vol. 7, No.3, pp. 276-296.

Adeli, H. and Kamal, O. (1990), "Automatic Partitioning of Frame Structures for Concurrent Processing," *Microcomputers in Civil Engineering*, Vol. 5, No. 4, pp. 269-283.

213

Adeli, H., and Kamal, O. (1992a), "Concurrent Optimization of Large Structures–Part I Algorithms," *Journal of Aerospace Engineering*, ASCE, Vol. 5, No. 1, pp. 79-90.

Adeli, H., and Kamal, O. (1992b), "Concurrent Optimization of Large Structures–Part II Applications," *Journal of Aerospace Engineering*, ASCE, Vol. 5, No. 1, pp. 91-110.

Adeli, H., and Kamal, O. (1993), *Parallel Processing in Structural Engineering*, Elsevier Applied Science, London.

Adeli, H., Kamat, M., Kulkarni, G. and Vanluchene, D. (1993), "Review of High Performance Computing Methods in Structural Mechanics," *Journal of Aerospace Engineering*, ASCE, Vol 6, No. 3, pp. 249-267.

Adeli, H. and Kao, W. M. (1996), "Object-Oriented Blackboard Models for Integrated Design of Steel Structures," *Computers and Structures*, Vol. 61, No. 3, pp. 545-561.

Adeli, H. and Kumar, S. (1995a), "Distributed Finite Element Analysis on a Network of Workstations - Algorithms," *Journal of Structural Engineering*, ASCE, Vol. 121, No. 10, pp. 1448-1459.

Adeli, H. and Kumar, S. (1995b), "Distributed Genetic Algorithm for Structural Optimization," *Journal of Aerospace Engineering*, ASCE, Vol. 8, No. 3, pp. 156-163.

Adeli, H. and Kumar, S. (1995c), "Concurrent Structural Optimization on a Massively Parallel Supercomputer," *Journal of Structural Engineering*, ASCE, Vol. 121, No. 11, pp. 1588-1597.

Adeli, H. and Yu, G. (1995), "An Integrated Computing Environment for Solution of Complex Engineering Problems Using Object-Oriented Programming Paradigm and a Blackboard Architecture," *Computers and Structures*, Vol. 54, No. 2, pp. 255-265.

AISC (1994), *Manual of Steel Construction-Load and Resistance Factor Design*, American Institute of Steel Construction, Chicago.

AISC (1994), *Manual of Steel Construction-Load and Resistance Factor Design - Vol. I - Structural Members, Specifications, and Codes*, American Institute of Steel Construction, Chicago.

AISC (1989), *Manual of Steel Construction-Allowable Steel Design*, 9th Ed., American Institute of Steel Construction, Chicago.

Ajiz, M.A. and Jennings, A. (1984), "A Robust Incomplete Cholesky-Conjugate Gradient Algorithm," *International Journal of Numerical methods in Engineering*, Vol. 20, pp. 949-966.

Al-Nasra, M. and Nguyen, D.T. (1991), "An Algorithm for Domain Decomposition in Finite Element Analysis," *Computers and Structures*, Vol. 39., No. 3, pp. 277-289.

Ashcraft, C.C., Grimes, R.G., Lewis, J.G., Peyton, B.W. and Simon, H.D. (1987), "Progress in Sparse Matrix Methods for Large Scale Linear Systems on Vector Supercomputers," *International Journal of Supercomputer Applications*, Vol. 1, No. 4, pp. 10-30.

Ayakanat, C., Ozguner, F., Ercal, F. and Sadayappan, P. (1988), "Iterative Algorithms for Solution of Large Sparse Systems of Linear Equations on Hypercubes," *IEEE Trans. on Computers*, Vol. 37, No. 12, pp. 1554-1568.

Barrett, R., Berry, M., Chan, T., Demmel, J., Donato, J., Dongarra, J., Eijkhot, V., Pozo, R., Romine, C. and Vorst, H. V. D, (1993), *Templates for the Solution of Linear Systems: Building Blocks for Iterative Methods*, SIAM, Philadelphia, NY.

Bathe, K. J. and Wilson, E. (1982) *Finite Element Procedures in Engineering Analysis,* Prentice Hall, New Jersey.

Belytchko, T., Plakacz, E. J., and Kennedy, J. M. (1990), "Finite Element Analysis on the Connection Machine," *Computer Methods in Applied Mechanics and Engineering*, Vol. 81, pp. 229-546.

Bianchini, R. and Brown, C. (1993), "Parallel Genetic Algorithms on Distributed Memory Architectures," Technical Report 436, Computer Science Department, The University of Rochester, NY.

Carey, G. F., Barragy, E., McLay R. and Sharma, M. (1988), "Element-by-element Vector and Parallel Computations," *Communication in Applied Numerical Methods*, Vol. 4, pp. 299-307.

Carey, G. F. and Jiang, B. (1986), "Element-By-Element Linear and Nonlinear Solution Schemes," *Communication in Applied Numerical Methods*, Vol. 2, pp. 145-153.

Chuang, L. C. and Adeli, H. (1993), "Design-independent CAD Window System Using the Object-oriented Paradigm and HP X Widget Environment," *Computers & Structures*, Vol. 48, No. 3, pp. 433-440.

Dongarra, J., Geist, G.A., Manchek, R. and Sunderam, V.S. (1993), "Integrated PVM Framework Supports Heterogeneous Network Computing," *Computers in Physics*, Vol. 7, No. 2, pp. 167-175.

Duff, I.S., Reid, J.K. (1983), "The Multifrontal Solution of Indefinite Sparse Symmetric Linear Equation," *ACM Transactions on Mathematical Software*, Vol. 9, pp. 302-325.

Duff, I.S., Erisman, A.M. and Reid, J.K. (1986), *Direct Methods for Sparse Matrices*, Clarendon Press, Oxford.

Duke, D. W. (1993), "Cluster Computing Exploits Performance and Cost Advantages," *Computers in Physics*, Vol. 7, No. 2, pp. 177-183.

Farhat, C., (1988), "A Simple and Efficient Automatic FEM Domain Decomposer," *Computers and Structures*, Vol. 28, No. 5, pp. 579-602.

Farhat, C., Sobh, N., and Park, K. C. (1989), "Dynamic Finite Element Simulations on the Connection Machine," *International Journal of High Speed Computing*, Vol. 1, pp. 289-302.

Farhat, C. and Wilson, E. (1988), "A Parallel Active Column Solver," *Computers and Structures*, Vol. 28, No. 2, pp. 289-304.

Farley, J. and Farley, J. (1998) *Java Distributed Computing*, O' Reilley & Assoc., Cambridge, MA.

Feit, S. (1996), *TCP/IP: Architecture, Protocols, and Implementations with Ipv6 and IP Security, 2nd ed.*, McGraw-Hill Book Company, New York, NY.

Fletcher, R. (1975), "An Ideal Penalty Function for Constrained Optimization," *Journal of Institute of Mathematics and Its Applications*, Vol. 15, No. 3, pp. 319-342.

Geist, G.A., Beguelin, A., Dongarra, J., Jiang, W., Manchek, R. and Sunderam, V. S. (1993), *PVM User's Guide and Reference Manual*, Technical Report ORNL/TM-12187, Engineering Physics and Mathematics Division, Oak Ridge National Laboratory, Oak Ridge, TN.

Goldberg, D. E. (1989), *Genetic Algorithm in Search, Optimization and Machine Learning.* Addison-Wesley Publishing Company, Inc., New York, NY.

Golub, G. H. and Loan, C. F. V. (1989), *Matrix Computations, 2nd Ed.*, The Johns Hopkins University Press, Baltimore, MD.

Grimes, R. T. (1997) *Professional DCOM Programming*, Wrox Press Inc., Chicago, IL.

Hartman, J. and Wernecke, J. (1996), *The VRML 2.0 Handbook*, Addison-Wesley Publishing Company, Reading, MA.

Hinton, E. and Campbell, J. S. (1974), "Local and Global Smoothening of Discontinuous Finite Element Functions Using a Least Square Method," *International Journal of Numerical Methods in Engineering*, Vol. 8, pp. 461-480.

Hoffmeister, F. (1991), "Scalable Parallelism by Evolutionary Algorithms," in Garuer, M. and Pressmar, D. B., Eds., *Parallel Computing and Mathematical Optimization*, Springer-Verlag, FRG.

Hsu, H. L. and Adeli, H. (1991) "A Microtasking Algorithm for Optimization of Structures," *International Journal of Supercomputer Applications*, Vol. 5, No. 2, pp. 81-90.

Irons, B. M. (1970) "A Frontal Solution Program," *International Journal of Numerical Methods in Engineering*, Vol. 2, pp. 5-32.

Janssen, T. L. (1983), "A Simple Efficient Hidden Line Algorithm," *Computers & Structures*, Vol. 17, No. 4, pp. 563-571.

Jennings, A. and Malik, G. M. (1978), "The Solution of Sparse Linear Equations By the Conjugate Gradient Method," *International Journal of Numerical Methods in Engineering*, Vol. 12, pp. 141-158.

Johan, Z., Mathur, K. K., Johnsson, S. L., and Hughes, T. J. R. (1994), "An Efficient Communication Strategy for Finite Element Methods on the Connection Machine CM-5 System," *Computer Methods in Applied Mechanics and Engineering*, Vol. 113, pp. 363-387.

Johan, Z., Hughes, T. J. R., Mathur, K. K., and Johnsson, S. L. (1992), "A Data Parallel Finite Element Method for Computational Fluid Dynamics on the Connection Machine System," *Computer Methods in Applied Mechanics and Engineering*, Vol. 99, pp. 113-134.

Johnsson, S. L. and Mathur, K. K. (1990), "Data Structure and Algorithms for the Finite Element Method on Data Parallel Supercomputer," *International Journal of Numerical Methods in Engineering*, Vol. 29, pp. 881-908.

Johnsson, S. L and Mathur, K.K, (1989), "Experience with the Conjugate Gradient Method for Stress Analysis on a Data Parallel Supercomputer," *International Journal of Numerical Methods in Engineering*, Vol. 27, pp. 523-546.

Khot, N.S. and Berke, L. (1984), "Structural Optimization Using Optimality Criteria Methods," in Atrek, E., Gallaghar, R.H., Ragsdell, K. M. and Zienkiewicz, O.C., Eds., *New Directions in Optimum Structural Design*, John Wiley & Sons, New York, NY.

Kao, W. M. and Adeli, H. (1997), "Distributed Object-Oriented Blackboard Model for Integrated Design of Steel Structures," *Microscomputers in Civil Engineering*, Vol. No. 12, No. 2, pp. 141-154.

King, R. B. and Sonnad, V. (1987), "Implementation of an Element-by-Element Solution Algorithm for the Finite Element Method on a Coarse-Grained Parallel Computer," *Computer Methods in Applied Mechanics and Engineering*, Vol. 65, pp. 47-59.

Kumar, S. and Adeli, H. (1995a), "Distributed Finite Element Analysis on a Network of Workstations - Implementation and Application," *Journal of Structural Engineering*, ASCE, Vol. 121, No. 10, pp. 1456-1462.

Kumar, S. and Adeli, H. (1995b), "Minimum Weight Design of Large Structures on a Network of Workstations," *Microcomputers in Civil Engineering*, Vol. 10, No. 6, pp. 423-432.

Kumar, S. and Adeli, H. (1997), "Distributed Finite Element Analysis on a Network of Workstations - Closure," *Journal of Structural Engineering*, ASCE, Vol. 123, No. 3, pp. 378-381.

Lesoinne, M. and Farhat, C. (1991), "Parallel/Vector Improvements of Frontal Method," *International Journal of Numerical Methods in Engineering*, Vol. 32, pp. 1267-1281.

Malone J. G. (1988), "Automated Mesh Decomposition and Concurrent Finite Element Analysis for Hypercube Multiprocessor Computers," *Computer Methods in Applied Mechanics and Engineering*, Vol 70, pp. 27-58.

Mandel, J. (1993), "Adaptive Iterative Solvers in Finite Elements," in Papadeakakis, M., Ed., *Solving Large-scale Problems in Mechanics*, John Wiley & Sons. pp. 65-88.

Mikes, S. (1990), *X Window System: Technical Reference*, Addison-Wesley Publishing Co, Reading, MA.

Noor, A.K. (1987), "Parallel Processing in Finite Element Structural Analysis," in Noor, A.K., Ed., *Parallel Computations and Their Impact on Mechanics*, American Society of Mechanical Engineers, New York, pp. 253-277.

Nour-Omid, B. and Park, K. C. (1987), "Solving Structural Mechanics Problem on the Hypercube Machine," *Computer Methods in Applied Mechanics and Engineering*, Vol 61, pp. 161-176.

Orfali, R. and Harkey, D. (1997), *Client/Server Programming with Java and CORBA*, John Wiley & Sons, Inc., New York, NY.

Pettey, C. C., Leuze, M. R. and Grefenstette, J. J. (1987), "Genetic Algorithms on a Hypercube Multiprocessor," in Heath, M. T., Ed., *Hypercube Multiprocessors 1987*, SIAM, Philadelphia, pp. 333-341.

Powell, M.J.D. (1969), "A Method for Nonlinear Constraints in Minimization Problems," in Fletcher, R., Ed., *Optimization*, Academic Press, London.

Quinn, M. J. (1993), *Parallel Computing: Theory and Practice*, McGraw-Hill Book Company, New York, NY.

Rank, E. and Babuska, I. (1987), "An Expert System for Optimal Mesh Design in the *hp*-Version of the Finite Element Method," *International Journal of Numerical Methods in Engineering*, Vol. 24, pp. 2087-2106.

Saad, Y. (1985), "Practical Use of Polynomial Preconditioning for the Conjugate Gradient Method," *SIAM J. Sci. Stat. Comput.*, Vol. 6, No. 4, pp. 865-881.

Saleh, A. and Adeli, H. (1994a), "Microtasking, Macrotasking and Autotasking for Optimization of Structures," *Journal of Aerospace Engineering*, ASCE, Vol. 7, No. 2. pp. 156-174.

Saleh, A. and Adeli, H. (1994b), "Parallel Algorithms for Integrated Structural and Control Optimization," *Journal of Aerospace Engineering*, ASCE, Vol. 7, No. 3. pp. 297-314.

Saleh, A. and Adeli, H. (1996), "Parallel Eigenvalue Algorithms for Large-Scale Control-Optimization Problems," *Journal of Aerospace Engineering*, ASCE, Vol. 9, No. 3, pp. 70-79.

Sarma, K. C. and Adeli, H. (1995), "Effect of General Sparse Matrix Algorithm on Optimization of Sparse Structure," *AIAA Journal*, Vol. 33, No. 12, pp. 2442-2444.

Sarma, K. C. and Adeli, H. (1996), "Sparse Matrix Algorithm for Minimum Weight Design of Large Structure," *Engineering Optimization*, Vol. 27, pp. 65-85.

Siegel, J. (1996), *CORBA: Fundamentals and Programming*, John Wiley & Sons, Inc., New York, NY.

Stroustrup, B. (1991), *The C++ Programming Language, 2nd Ed.*, Addison-Wesley Publishing Co., Reading, MA.

Suarjana, M. (1996), "Comparison of Preconditioned Conjugate Gradient Method and LDL Factorization for Structural Analysis," *Computer-Aided Civil and Infrastructure Engineering*, Vol. 13, No. 4, pp. 289-296.

Thinking Machines (1991), *CM-5 Technical Summary*, Thinking Machines Corporations, Cambridge, MA.

Thinking Machines (1992a), *The Connection Machine CM-5 Technical Summary*, Thinking Machines Corporation, Cambridge, MA.

Thinking Machines (1992b), *CM Fortran Reference Manual, Version 2.0 Beta*, Thinking Machines Corporation, Cambridge, MA.

Thinking Machines (1993a), *CMMD Reference Manual, Version 3.0*, Thinking Machines Corporation, Cambridge, MA.

Thinking Machines (1993b), *CMMD User's Guide, Version 3.0*, Thinking Machines Corporation, Cambridge, MA.

Thinking Machines (1993c), *CM Fortran Utility Library Reference Manual, Version 2.0 Beta*, Thinking Machines Corporation, Cambridge, MA.

Thinking Machines (1993d), *CMSSL for CM Fortran: CM-5 Edition, Version 3.1 Beta 2*, Thinking Machines Corporation, Cambridge, MA.

Turcotte, L. H. (1993), "A Survey of Software Environments for Exploiting Networked Computing Resources," Technical Report, Engineering Research Center for Computational Field Simulation, Mississippi State, MS.

White, D.W., and Abel, J.F. (1988), "Bibliography on Finite Elements and Supercomputing," *Communications in Applied Numerical Methods*, Vol. 4, No 2, pp. 279-294.

Winget, J. M. and Hughes, T.J.R, (1985), "Solution Algorithms for Nonlinear Transient Heat Conduction Analysis Employing Element-by-element Iterative Strategies," *Computer Methods in Applied Mechanics and Engineering*, Vol. 52, pp. 711-815.

UBC (1994), *Uniform Building Code*, International Conference of Building Officials, Whittier, CA.

Vogel, A. and Duddy, K. (1997), Java Programming with CORBA, John Wiley & Sons, Inc., New York, NY.

Yu, G. and Adeli, H. (1993), "Object-oriented Finite Element Analysis Using EER Model," *Journal of Structural Engineering*, ASCE, Vol. 119, No. 9, pp. 2763-2781.

Zienkiewicz, O. C. and Zhu, J. Z. (1987), "A Simple Efficient Error Estimator and Adaptive Procedure for Practical Engineering Analysis," *International Journal of Numerical Methods in Engineering*, Vol. 24, pp. 337-357.

Author Index

A

Abel, J. F., 6
Adeli, H., 5, 6, 24, 31, 47, 57, 87, 88, 89, 90, 93, 95, 97, 126, 134, 135, 136, 137, 147, 148, 176, 177, 203
AISC, 87, 92, 169, 127
Ajiz, M. A., 204
Al-Nasra, M., 24
Ashcraft, C. C., 202
Ayakanat, C., 42

B

Babuska, I., 5, 202
Barrett, R., 6
Bathe, K. J., 185, 186, 188
Belytchko, T., 148
Berke, L., 88, 123, 125, 126
Bianchini, R., 106
Brown, C., 106

C

Campbell, J. S, 9
Carey, G. F., 11
Cheng, N. T., 31, 88, 89, 90, 93, 95, 134, 135, 136, 137, 148, 176, 177
Chuang, L. C., 57

Subject Index

T - #0125 - 101024 - C0 - 234/156/14 [16] - CB - 9780849320934 - Gloss Lamination